T0253615

Gaining the High Ground over Evolutionism

Gaining the High Ground over Evolutionism

ROBERT J. O'KEEFE

Gaining the High Ground over Evolutionism

Copyright © 2012 by Robert J. O'Keefe

All rights reserved. No part of this book may be used or reproduced by any means, graphic, electronic, or mechanical, including photocopying, recording, taping or by any information storage retrieval system without the written permission of the author except in the case of brief quotations embodied in critical articles and reviews.

iUniverse books may be ordered through booksellers or by contacting:

iUniverse
1663 Liberty Drive
Bloomington, IN 47403
www.iuniverse.com
1-800-Authors (1-800-288-4677)

Because of the dynamic nature of the Internet, any web addresses or links contained in this book may have changed since publication and may no longer be valid. The views expressed in this work are solely those of the author and do not necessarily reflect the views of the publisher, and the publisher hereby disclaims any responsibility for them.

Any people depicted in stock imagery provided by Thinkstock are models, and such images are being used for illustrative purposes only.

Certain stock imagery © Thinkstock.

ISBN: 978-1-4759-4962-9 (sc)
ISBN: 978-1-4759-4964-3 (hc)
ISBN: 978-1-4759-4963-6 (e)

Library of Congress Control Number: 2012916876

Print information available on the last page.

iUniverse rev. date: 01/07/2016

Contents

Preface ix

Introduction xiii

Chapter 1: Science 1 1

Chapter 2: Genesis 17

Chapter 3: Geology 27

Chapter 4: Biology 41

Chapter 5: Astronomy and Physics 53

Chapter 6: Science 2 81

Chapter 7: Controversy 87

Chapter 8: Thought 113

Chapter 9: Philosophy 1 .123

Chapter 10: Philosophy 2 137

Chapter 11: Revelation .151

Chapter 12: Reason and Faith 155

Summary and Conclusion 163

Appendix A: Randomness or Natural Law? 167

Appendix B: Remarks Pertaining to the Validation of Premises 171

Notes 173

Bibliography 191

NIV scripture quotations from the Holy Bible, New International Version™ Copyright © 1973, 1978, 1984 by International Bible Society. Used by permission of Zondervan. All rights reserved.

RSV scripture quotations from The Catholic Edition of the Revised Standard Version of the Bible. Copyright © 1965, 1966 by National Council of the Churches of Christ in the United States of America. Used by permission. All rights reserved.

The quotation from Donald Goldsmith in chapter 6 reprinted by permission of the publisher: *Einstein's Greatest Blunder? The Cosmological Constant and Other Fudge Factors in the Physics of the Universe* by Donald Goldsmith, Harvard University Press, Cambridge, MA, Copyright © 1991 by the President and Fellows of Harvard College.

Preface

I like to think, and there is historical precedence for this, that there are three levels of the mind that apply to belief systems. Interaction with the world is the first level. It consists of experience or, if you prefer scientific ways of thinking, evidence. Experience or evidence determine the content of the second level, that of philosophy or the belief system itself. The third is that of will, that which, given a belief system, decides whether or not to be guided accordingly. I do not suppose anyone perceives philosophy to be an underconstrained problem in that any one of a number of belief systems is likely to work as well as another when one is interacting with the world. But there is reason to think that the process can run in reverse. A belief system may be picked according to what one deems *ought* to be found in experience, or it may be imposed by culture or religion. That belief system then rules how experience or evidence is interpreted or, as may be the case, misinterpreted.

It is quite possible to ignore evidence that does not fit one's philosophy, or hold that evidence in a state of unconsciousness. Implications of that evidence may be operative in practice, though not constraining on the belief system. Or the belief system may function as a sort of fortress, protecting the will level from having to deal consciously with evidence (or experience) that it does not see as a good fit to a preferred reality. An education program is responsible for the philosophy level, the belief system. Education should see that such a condition as a mismatch between experience and belief is not produced. If it is produced, it would be fair to ask why. Given my own experience, my belief is that any worthwhile education program should critique prevailing belief systems in the culture at large and, where found deficient, step toward a system that better aligns with experience. It would seem appropriate that such teaching should become a standard part of the secondary education curricula. Hence, this book along with the accompanying workbook is intended as an educational resource for late secondary school education. Of course, a text with such a title as this one could not be approved for use in a public school classroom. But public schools are not the only places where education proceeds.

Among the hazards of writing on a subject as wide and deep as the origin of the universe, the earth, and all living things is the impracticality of covering every aspect to the satisfaction of everyone. In answer to possible complaints, there is already much written on the controversy at hand that does not need to be reiterated. Much of this is found within the scientific sphere and is at the level of experience and evidence. The focal point here rests more in recognition of a certain complexity in the interaction between experience and belief. It is apparent that throwing more evidence at a controversy does not often result in its resolution or convince anyone not already predisposed to accept the evidence. It should be more constructive, if at all possible, to deal with a controversy at the second level. The third level is, of course, off limits. The controversy at hand is, therefore, addressed less as a question of scientific evidence and more as a philosophical problem with a history—how what began as a stipulation of scientific inquiry ended up as a conclusion.

I have found it necessary to draw out of the literature those aspects that seem to me the more relevant and perhaps underappreciated. The following is a chapter-by-chapter overview: Chapter 1 covers the scientific revolution and certain of its effects on Western thought. Chapter 2 covers the interpretation of Genesis 1. Chapters 3, 4, and 5 describe relevant aspects of the history of geology, biology, and astronomy. The more time-conscious reader or teacher may wish to skip over the history and relativity theory sections of chapter 5 for expediency sake. I have included these sections because that material should lead to greater appreciation of how science has worked and continues to work today. Chapter 6 summarizes science as a means of gaining knowledge, discusses the scientific method as it is applied to natural history, and then introduces the second half of the book. Chapter 7 reviews the history of how the court system has dealt with the controversy provoked by this subject in science education. In chapter 8, an attempt is made to draw points from C. S. Lewis's argument against naturalism (from the third chapter in his book *Miracles*, which is titled "The Cardinal Difficulty of Naturalism") and put them into a more structured form that hopefully may be taught more easily. Some thoughts of my own are added. Chapters 9 and 10 confront the ideology behind evolutionary science, the philosophy of naturalism, and present what I believe are the best arguments against it. Chapter 11 recalls the grounds for the authority of the Bible. Finally, chapter 12 discusses the necessary partnership of reason and faith.

Introduction

*The first to present his case seems right, till another
comes forward and questions him.*
—Proverbs 18:17 NIV

The editorial section of a local newspaper often contains examples of how people think about difficult and controversial subjects. The origin of the universe, earth, and living things is one of these subjects. The below excerpts of two letters to the editor defend the scientific view.

One writer expounds, "Creationism is doing a major disservice to our children and is causing much confusion. It is built on ignorance and misunderstanding of the biblical message and, if pursued, will lead us back to the Dark Ages."[1] He then states, "The Bible ... is a living book. Our understanding of it changes from generation to generation." If our understanding of the Bible changes from generation to generation, are there any grounds for saying that someone else's understanding is a misunderstanding? He writes that science and the Bible teach us different lessons: "The Bible seeks to answer the question of why the world was created and what is the meaning of life. Science answers the question of how Earth and the universe were formed and how life came into being." One might wonder that the whys and hows can be so easily disentangled. After writing that "Earth in the Bible is described as flat," he concludes, "We need to have equal respect for faith and science, because both can help us make a better world." How does he expect that anyone would have respect for faith after having just attributed (falsely) such a mistruth to faith's object? Is faith really as unconcerned with truth as he would lead us to believe?

Another opinion holder, a physician, referring to the intelligent design movement writes, "The argument that we are too complex to have evolved by chance is without basis."[2] He then goes on to state, "However, evolution is not based on mere chance. By following the principles of physics and biology, nature has forced our evolution through environmental pressures. With each step of evolution, less and less randomness was needed. It can be argued that if the world was restarted over and over again, the same evolutionary steps would be followed if the same environmental pressures are present." It is puzzling that one proposition (intelligent design) can be dismissed by appeal to the power of chance yet that the opposing proposition (evolution) is not based on chance either, particularly since the same evolutionary steps would be followed each time. Diseases are next offered as positive proof that an intelligent designer does not exist. There is a hidden assumption that the goal of a supernatural being must be nothing short of physical perfection. He ends by writing that despite the wondrous and confusing mystery the human anatomy and physiology presents to us, "we need to stop imposing the supernatural to fill our knowledge gaps and keep using that most amazing biological instrument, our minds." Does not this argument make his mind, and therefore his assertions, an inevitable outcome of principles of physics and biology?

The newspaper's editors weigh in, drawing a sharp distinction between science and religion: "Intelligent design clearly should not be taught in science classes as an alternative to the theory of evolution. It's not science; it's a belief rooted in creationism."[3] In contrast, "Evolution ... has been thoroughly investigated and has mountains of evidence supporting it." They continue, "Intelligent design, the idea that life on Earth is too complex to have evolved naturally and therefore must have been designed by a supernatural creator, cannot be proved or disproved. So it isn't science." They go on to state that religious beliefs, including intelligent design, are faith-based, unprovable, and that as such, they are controversial ideas. They also mention "the battle over the proper place of religion in public life" and how students may be "enriched by discussions of the effects that belief has on politics and culture." Meanwhile, evolution seems safely tucked away within the category of science, outside the realm of controversial ideas, and free from public scrutiny of a similarly unprovable idea that supernatural causes were not involved.

Appeals to popular views and dispositions with some occasional not-too-rigorous logic are not what one would expect to find being used to defend scientific theories. These are more characteristic of debates on political questions and ideological beliefs. Yet debates surrounding ideas about the origin of things may not be as much a controversy over scientific theory as it is wrangling over ideology. As much as scientists regard evolution as existing within the confines of scientific theory, the influence and ramifications of evolutionary theory extend beyond the scientific profession. Evolutionary thinking affects more than science in today's world, if only because your sense of where you came from informs your sense of meaning and value and because your sense of meaning and value informs everything else. The vacuity of scientific answers in the realm of meaning and value has often been interpreted as grounds for belief that there is no meaning or value. Interestingly, that is not how people act; they act as if life really does have meaning and really does have value, especially when it ends. Why, even science has meaning and value if it improves the quality of that which is thought to have meaning and value. The questions about meaning and value (or lack of it) raised by evolutionary theory may have something to do with why nonscientists take such an interest in its standing.

Creation exists as the sole counter-perspective because, at a certain level, the options are reduced to two: natural and supernatural. But, we are assured, evolution is a scientific idea whereas creation is a religious idea. We are further assisted in our mental categories by professional philosophers and scientists who have defined for us the content of science and religion. Science is reasoned, objective, and factual. The scientific method is reliable, self-correcting, and impartial. Religion, we are further taught, does not require or even accept the burdens of logic and evidence. So the number of viable options is down to one. Indeed, beginning with the Enlightenment period of history in the eighteenth century, Western culture has sought to enshrine the methods of science as *the* foundation of knowledge. The scientific disciplines have informed our understanding of the natural world and delivered the knowledge from which great gains have been made in improving the quality of life in so many respects. The esteem given the scientific approach to solving the puzzles of the universe appears justified. Yet the fact that scientific theory impinges on subject matter it is ill-equipped to handle and the nonscientific manner of its defense observed above leaves us with grounds for suspicion: Is science being used to serve some other agenda?

Chapter 1:
Science 1

It requires a very unusual mind
to undertake the analysis of the obvious.
—Alfred North Whitehead

Yes, science has enabled us to understand the natural world, how it operates, and how to exploit that knowledge to our advantage. It has also produced a comprehension of how the natural world came into existence. Because science is so often at the center of controversy over the origin of the universe, earth, and all living things, it is worthwhile to take a look at its history. Science has not always been the lens through which the natural world is known. Among historians, the beginning of modern science has come to be known as the scientific revolution, not because it involved wars and assassinations but because it produced such a revolutionary transformation of Western thought. That transformation of thought is the topic of interest in this chapter.

Historians generally place the scientific revolution within the seventeenth century. It began to emerge shortly after the Renaissance, that period of European history occupying the late fifteenth and early sixteenth centuries, typically marking the end of the Middle Ages and the beginning of modern ages.

WHAT SET THE STAGE

The Renaissance was characterized by a notable increase in the rate of discovery, creativity, and invention. It inspired an accelerated rate of learning and accumulation of knowledge that has continued to the present day. The Age of Discovery and the Protestant Reformation were occurring during that time. The Age of Discovery refers to when mariners began to chart and navigate trade routes from Europe to India, China, and the East. Navigation on the open seas became practical once the Chinese invention of the magnetic compass was improved to the point of being a reliable navigation instrument. Exploration with intent to find an alternate route to the East led to the first circumnavigation of the globe in 1521 by Ferdinand Magellan (1480–1521).[1] Advances in geographical reach accompanying commerce and exploration brought an expanding knowledge of the world and an increasing awareness of other civilizations.

When Martin Luther (1483–1546) attached ninety-five theses to the door of a church building in Germany in 1517, he inadvertently kicked off another revolution of sorts known as the Protestant Reformation. The Counter-Reformation, Roman Catholicism's reply to the various Protestant movements that broke out, occupied the second half of the sixteenth century. Fueled by economic and political agendas, the Reformation and Counter-Reformation eventually led to a great deal of religious strife, sectarian conflict, and religious wars, the Thirty Years' War (1618–48) in particular. An irreversible fragmentation of what was perceived to be religious authority in Western Europe had occurred.

The Chinese inventions of paper and printing led to the invention of moveable type in about the year 1450. The consequent mass production of books in common languages meant that knowledge was no longer the exclusive domain of a select few schooled in Latin. Besides the expansion of knowledge, the Renaissance period is also noted for achievements in art, music, architecture, and literature. Moreover, the past preoccupation with theology and the providential acts of God toward mankind was beginning to be replaced with things mankind could do in service to mankind. This trend is known among historians as Renaissance humanism.[2] These factors—exploration of the world, fragmentation of religious authority, widespread dissemination of knowledge, and the increasing realization of mankind's creative and intellectual capabilities—as well as many other factors not mentioned made for the environment in which the scientific revolution began to take root.

If the immediate historical context of the scientific revolution contained factors influential to its beginning, other earlier developments, particularly in the twelfth and thirteenth centuries, could be said to be at least as important, if not more so. The writings of the ancient Greek philosophers

Plato and Aristotle were translated from Greek to Arabic in the eighth century, and copies were eventually brought to Western Europe. The ancient texts were retranslated from Arabic to Latin, and later, they were translated directly to Latin from the Greek manuscripts. They inspired a pre-Renaissance renaissance in learning during the twelfth and thirteenth centuries, and universities devoted to the study of these texts were founded at that time.

Accompanying the infusion of ancient Greek philosophy, a fundamental advancement in reasoning helped to free knowledge from reliance on received traditions. This advancement was inspired by the manner of the ancient Greek philosopher Socrates. Socrates went about challenging commonly held beliefs and assumptions by asking questions intended to reveal logical consequences that often showed such beliefs and assumptions to be faulty. Socrates's tactic is generally known as the dialectic method of reasoning. In the late Middle Ages, an adaptation of the dialectic method subjected knowledge to objective testing by demanding consideration of both arguments and counterarguments for a given proposition.[3] A proposition could be sustained or defeated by the strength of an argument, or a synthesis could be formed wherein apparent contradictions between opposing propositions would be reconciled, the synthesis often leading to new knowledge. Confidence in the certainty of knowledge was increased when further grounds for it could be found beyond solely the authority of a source. This more systematic approach to knowledge was being applied to disputes over matters of law, philosophy, and theology in the late Middle Ages. Philosophical and theological inquiries were also granted a measure of independence, at least in principle, from the pressure to serve religious and political objectives. The universities were places where such inquiry and associated discourse could proceed autonomously from church and state authorities.

Various fields of study inherited from the ancient Greeks and revived in the universities of the late Middle Ages included several readily recognizable branches of science: astronomy, chemistry, physics, optics, anatomy, and physiology. However, we must realize that science was not a distinctly recognized category of thought until the nineteenth century. Scientists were not scientists but natural philosophers; science was not called science but natural philosophy. That which sought to describe and explain the natural world was regarded as a philosophical pursuit. It was not the same as the practice of science we know today. Natural philosophy was not far from natural theology, which sought to learn about God from nature and from whatever inherent knowledge of God might exist already in the mind of man. For all practical purposes, during the time period of the scientific revolution, natural theology and natural philosophy were synonymous terms.

Modern Science Arrives

The world upon which the Renaissance dawned had inherited the ancient Greek philosophy of Aristotle (384–322 BC). Natural philosophy was inspired by the legacy of Aristotle, and Aristotle applied logic to understand the nature of things. But his philosophy, while rational, was influenced more by ideals than reality; he was not concerned with observations to the degree of incisiveness demanded in science today, and neither was he concerned with experiments or mathematical descriptions. This is why natural philosophy was not the same as today's science. It was neither empirical nor mathematical. Aristotle thought in terms of there being four causes in operation in the world:

- Material cause of composition (i.e., a house is made of wood, stone, and other materials).
- Formal cause of form, design, or plan (i.e., a house's blueprint).
- Immediate cause as a precipitating event (i.e., the builders).
- Final cause of purpose or what use something was intended for (i.e., a house as a building to dwell in).

But of Aristotle's four categories of causes, only immediate cause (i.e., the builders) was retained by the natural philosophers. In the natural world, these immediate causes were various forms of energy (or forces) acting upon objects. Though unseen, the existence of such forces was predicated on the ancient Greek philosophical idea that the natural world is ruled by a rational ordering of cause and effect, that every effect is a consequence of preceding causes. The emphasis on immediate cause and the exclusion of purpose and design became known as the mechanical philosophy. Presumably, narrowing the scope of inquiry in this manner was a consequence of the fact that the purpose and design of naturally occurring objects is not very discernible from the objects themselves. So the mechanical philosophy restricted the field of causative principles to the motions of objects, their actions and reactions. This trend in natural philosophy was most attributable to the philosophers Rene Descartes (1596–1650) and Thomas Hobbes (1588–1679).

Besides the mechanical philosophy, at least two other developments are seen by historians as essential to understanding the progress of science in the seventeenth century. The linkage of mathematics to natural phenomena was one. Mathematics had long existed, and so had natural philosophy, but that the two might be somehow connected was not a topic of consideration during the Middle Ages. Mathematics and natural philosophy were separate and unconnected disciplines in the universities. Mathematics was necessary for practical uses: constructing buildings, making mechanical devices like clocks, predicting the motion of planets, and in navigation, art, and music. Natural philosophy was concerned with form and function in the natural world. The separation between natural philosophy and mathematics was particularly apparent in how the motions of planets were described. The calculations that best described the motions of planets were not reconcilable with the way natural philosophers, inspired as they were by Aristotle, understood the ordering of the universe. Aristotle had imagined that the universe was formed of perfect spheres in perfectly circular orbits. But the observed motions of the planets were not at all circular. The mathematics of planetary motion was based on a complex artificial model of the universe that worked for the purposes of calculations. This model was inconsistent with the Aristotelian view, which was thought to be the correct view. That mathematics might describe the way the natural universe really worked was a radical idea. Nicolaus Copernicus (1473–1543) in his 1543 *De Revolutionibus Orbium Coelestrium* (*On the Revolutions of the Heavenly Spheres*) had first understood this—that mathematics must truly describe how the universe is configured. This idea led him to the realization that the arrangement of the known universe must be drastically different than what had always been assumed.

A second essential development was a requirement for observation and experiment to verify (in rigorous fashion) any claim to have obtained knowledge about the natural world. Experimentation is a subset of observation, a means to force nature to reveal characteristics that would not otherwise be observed. Francis Bacon (1561–1626) was the chief advocate of what has come to be known as the experimental or scientific method. As described in his 1620 *Novum Organum* (*The New Instrument*), the method was characterized by first the acquisition of facts from observations or experiments and then the generalization or induction from those facts to unifying concepts, natural laws, or theories about nature. This is the root of the ideal of the scientific method we know today. It was promoted as a way of ensuring the empirical grounding of knowledge (i.e., that knowledge was not just a clever scheme of the mind but that it indeed represented what was true about the natural world). Much like the adaptation of the dialectic method that subjected theological and philosophical propositions to rigorous evaluation, the experimental method made any knowledge about nature subject to rigorous verification. Bacon was highly critical of natural philosophy unsubstantiated by any empirical evidence. The experimental method also allowed for the resolution of disputes about facts of nature in a more civilized and rational manner than had often been used (and was still being used) to resolve many theological disputes.

The ideal of the experimental (or scientific) method has served well to advance the cultural authority of science. The method has been extended to include other steps involving the formulation of hypotheses and repeatable testing before hypotheses can be accepted as theories. But how strictly the ideal of the scientific method has been realized and applied in devising any particular theory of science is a matter that needs to be judged on a case-by-case basis. In contrast to the ideal of the scientific method, unifying philosophical ideas could also be and have been brought to collections of observed facts to explain and make sense of them. Unifying ideas or theories are not always derived exclusively from the facts to be explained. Uniformitarianism, an idea discussed later in this chapter, is an example of a unifying idea brought to the facts. The bringing in of other ideas as theories is actually more common than one might think. In fact, the ideal of deriving unifying concepts, theories, and laws exclusively from collections of facts is never entirely realized. This book is about an instance of other ideas brought in. We will return to discuss the scientific method in chapter 6.

In 1687, Isaac Newton (1642–1727) drove the golden spike that joined the empirical, mechanical, and mathematical trains of thought. This conjunction of thought was realized in his 1687 *Philosophiae Naturalis Principia Mathematica* (*Mathematical Principles of Natural Philosophy*). As an outcome of mathematical reasoning, he formulated three laws of motion: inertia, acceleration, and equal and opposite reaction, and a fourth law, the law of universal gravitation. Newton, in successfully describing these laws of motion and gravitation mathematically, is considered to have thereby completed the linkage between mathematics and natural phenomena understood by Copernicus, and further developed by Johannes Kepler (1571–1630), Galileo (1564–1642), and Rene Descartes.[4] Mathematically described and empirically verified by subsequent observation and successful prediction of the behavior of objects in motion, Newton's achievements sealed the gains of the scientific revolution.

At the same time that the field of inquiry into the natural world was being limited to immediate mechanical causes, there was a demand for empirical verification of knowledge about nature and a discovery that nature conforms to mathematical logic. These two developments within natural philosophy are what made "science" the science we know and value today. They were achieved over nearly a century and a half through the accumulated work of a number of individuals. But the transition that distanced science from natural theology also took time. It is a matter of historical record that occult influences on mechanical philosophy, which was then becoming science, and debate over mechanical philosophy's theological implications were substantial throughout the time span of the scientific revolution.

BUT HOW SCIENTIFIC WAS IT?

As strange as it may seem by today's understanding of science, the idea that supernatural forces operate within nature played a role in the development of science. Interest in the occult and in magical power was increasing during the Renaissance. Occult practices involved the manipulation of supernatural powers acting on the material world to achieve desired effects. The mechanical philosophy was concerned with understanding cause and effect in the natural world, but it was also concerned, just as occult practices were, with how such knowledge could be put into practical service by intervening in and exploiting the properties of objects.[5] The invisible forces exhibited by nature or "natural magic" were seriously studied by natural philosophers. As a prime example, magnets were considered objects with magical power because they had mysterious properties. Yet their properties were very predictable. Newton understood gravity as a type of natural magic.[6] Though forces and their effects in the natural world were at first attributed to the supernatural, they were eventually distanced from the supernatural and joined to the empirically verifiable and mathematically describable mechanical philosophy because of their observed consistencies.

In today's world, we do not think of natural forces as magical, but the difference between the magical view of natural phenomena and the cause-and-effect view of the mechanical philosophy is not great. We do not have any more of a fundamental understanding of forces of nature such as magnetism and gravity than natural philosophers did back then. We can measure, formulate, calculate, and predict cause-and-effect operations involving these forces, but we do not know why they are what they are. Why does gravity attract and not repel? We do not know. How can the presence of one object so affect another when there is no contact between them? We do not know. Why should there necessarily be any forces acting upon objects at all? We do not know.

A theological dispute that arose in reference to the mechanical philosophy shows how greatly theology and natural philosophy were intertwined. Newton thought that God arbitrarily established properties of matter without being restricted by any preexisting absolute properties. There were no rules that nature necessarily had to follow. He considered that whatever God willed was good. This thinking led to an empirical approach to knowledge, a position that knowledge of the natural world could only be obtained by observation. Conversely, according to Gottfreid Wilhelm Leibniz (1646–1716), a contemporary of Newton and also a mathematician and physicist, God was constrained to creating the world in accordance with preexisting absolute and inviolable properties of matter. Nature could only be made to exist under certain rules already governed by a transcendent rationality. He considered that God willed that which was objectively good. This thinking led to a position that knowledge of the natural world could only be discerned through reason, not observation.[7] The truths discovered by reason, it was thought, would be the same truths that God would be bound to. The dispute was an artificial conflict that forced a choice between making God's commands either arbitrary or else subject to a still higher authority. A synthesis between these opposing viewpoints could have been made by centering both rationality and goodness in God. The conflict was ultimately resolved some two centuries later by ejecting theology from science. Notwithstanding, there were obvious and substantial interactions between theology and natural philosophy in the early years of science.

IS THERE A PLACE YET RESERVED FOR DESIGN?

Natural philosophers saw themselves as "priests" of nature interpreting the "book" of nature. Prior to the nineteenth century, it was taken for granted that the two books, the book of nature and the book of divine revelation, the Bible, agreed. All this flurry of inquiry into the operations of the natural world remained culturally acceptable as long as whatever knowledge was obtained substantiated belief in the Creator. But the mechanical philosophy had a tendency to cause those viewing it from the outside to suspect that the new philosophy was drifting toward an atheistic view of the natural world. The emphasis on discovering comprehensive mechanical laws of nature tended to displace anything God might do to the point that the importance of God diminished toward irrelevance. The laws themselves became the source of the supreme order rather than God, the lawmaker. While natural philosophers sought to use natural philosophy to substantiate religious faith, they often found themselves arguing that their philosophy was not in fact undermining that faith.[8] Among natural philosophers arguing on behalf of faith in God and against the perceived drift toward atheism were Isaac Newton, Robert Boyle, Gottfried Leibniz, and Francis Bacon.

This point is intriguing. The argument being put forward by natural philosophers was that natural philosophy revealed design in nature, which pointed to the existence of a creator. This argument is known as the teleological argument, the argument for the existence of a creator from design or purpose. But if the inquiry and knowledge that was tending to lead to atheism was also being used to confirm theism—that is, the existence of a designer—might there have been something more fundamental that should have determined the conclusion? The mechanical philosophy had already excluded the consideration of purpose and design in nature. Excluding purpose and design was thereby a precondition, not a conclusion of scientific investigation. So whether or not a purpose and design existed in nature was perhaps not a question that scientific investigation alone could have answered.

The same argument (design in nature) continues to be used today in the intelligent design movement: certain scientific findings (the information content of DNA, the intricate complexities of living cells, etc.) lead to inferences that living things have been intelligently designed. But as the mechanical philosophy was then, the practice of science today is similarly predicated on excluding the consideration of purpose and design in nature. A charge that science is atheistic is therefore not without grounds when science attempts to explain our existence without reference to any purpose or design. The design argument did not stop the trend toward atheism then, and it is doubtful that it can turn science back toward theism today. This is because the answer to the question of whether or not the natural world is designed seems to have turned on an upfront decision on whether or not to consider design. What history shows is that science's answer depends not on scientific findings but on a predisposition to accept that idea or not.

Modern thinking is often confused: The precondition of science is being taken as a conclusion of science by many who would rather not have to deal with supernatural factors.

It is not that the design argument is ineffective; it can be persuasive, and it has been in use for centuries because it is persuasive.[9] But a more effective strategy would be to restore the full scope of inquiry from the beginning. Aristotle's other causes need to be readmitted upfront instead of attempting to bring them back in through a method of inquiry that already put them out. Thus, there is another wider view of the design argument that needs to be known and applied, that design absolutely exists in the world because of human intervention in nature. It is not necessary to rely on nature to prove the point. Any manmade object is proof that design exists in that man-made objects are instantly distinguishable from naturally occurring objects. They all have a definite purpose, a purpose that does not arise from natural causes even if the purpose is not immediately recognizable. Therefore, something outside of nature must exist. We will return to this subject in chapter 10.

A CONFLICT OF WHICH MUCH IS OFTEN MADE

Early in the seventeenth century, Copernican heliocentricity (sun-centered universe) began to seriously challenge everyone's understanding of how the universe was configured. Over a period of more than thirty years, Nicolaus Copernicus had toiled to perfect the argument for positioning the sun at the center of the universe. This he did on mathematical grounds alone, culminating in his completion of *On the Revolutions of the Heavenly Spheres* in 1543. But until Galileo (1564–1642) began to obtain certain visual confirmations via the newly invented telescope in 1609[10], his theory had been accepted more as a mathematical convenience rather than a physical truth. Copernicus himself was certain of the truth of heliocentricity, but very few others were convinced. While geocentricity (earth-centered universe) is not an explicit teaching of the Bible, everyone assumed it was. The Joshua 10 event where the sun remained still in the sky implied a moving sun and stationary earth. Truly, that reference is from the perspective of an earthbound observer (which everyone was). Nevertheless, geocentricity was so firmly established, so long enduring, and so intuitively obvious that it effectively functioned as a religious doctrine.

The confrontation between Galileo and the Roman Catholic Church is typically presented as a science-versus-religion conflict. But there are inconsistent historical accounts and indications that the confrontation may have been more complicated and more personal. The Catholic church was well aware of Copernicus's theory but had made no official pronouncements on it. Galileo thought he had definitive evidence of a sun-centered universe and was floating the idea in public. In 1616, he was formally advised by church officials to stop pressing the matter, which he did for a time. Then in 1632, he published a book appealing to popular audiences, one that ridiculed the earth-centered universe idea. The book was structured as a dialog between two characters: Simplicio, who argued in favor of an earth center, and Salviati, who argued in favor of the sun. It was initially titled *Dialogue Concerning the Tides* because Galileo thought tides were evidence of an earth in motion. Galileo was advised by the church's censors to change the title to *Dialogue Concerning the Two Chief World Systems.*[11] This undoubtedly was a more descriptive title, and by so advising, the church's censors actually approved his book! Then later, for reasons that remain unclear, Pope Urban VIII ordered the Holy Office (Inquisition) to prepare a case against Galileo. This took place the following year in 1633. In preparing its case, the church ruled officially on the "world system" question. It may be a relevant factor that Urban VIII had been a friend of Galileo when he was formerly known as Cardinal Barberini, and Galileo's choice of name for the character who argued in favor of the earth-centered viewpoint may have had something to do with his running afoul of the pope. The world system question was by no means settled among natural philosophers, and to suggest to the public that thoughtful opponents were simpletons may not have been the most tactful persuasive strategy. This time, Galileo was kept under house arrest until his death, and all books wherein a Copernican view of the universe was expounded, including that of Copernicus, ended up on the Index of Prohibited Books.

Scientific Thought Begins to Reign

In accordance with the trend in thinking brought about by natural (now mechanical) philosophy and Newton's success with it, and the overthrow of an established doctrine (geocentricity) by mathematical necessities, a major shift in how the world was understood was taking place. Miraculous intervention was out, and natural law was in. Divine revelation was out, and reason was in as the highest authority brought to bear in making sense out of the natural world. The Enlightenment, otherwise known as the Age of Reason, a period of history roughly coinciding with the eighteenth century, was the product of this major shift in thought. The theology or philosophy characteristic of this period is known as deism. Most generally, deism was the displacing of divine revelation with reason as the final instrument of truth. The popularity of deistic philosophy during this period was a significant historical stage between the theism of the Middle Ages and the full development of atheism in the twentieth century. The developments in thought presupposed a rational consistency that ruled the natural world. But the direction being taken would eventually lead to a loss of awareness that there was a cause for that rational consistency.

The one specific idea most associated with the philosophy of deism is that of the Creator establishing laws governing the operations of nature and then standing back from the created world and allowing it carry on by itself without any further involvement. This idea necessarily excludes any continuing miraculous or supernatural intervention in the operations of the created world. The laws of nature, once set in order, are considered supreme. That they would need to be interfered with from time to time was considered a crude, primitive theology, unworthy of a perfect creator. By the end of the eighteenth century, this philosophy was a common theological position and understanding of the way the world worked among the scientifically (philosophically) educated.

The supremacy of natural laws and the absence of supernatural intervention was precisely the philosophical environment within which scientific inquiry and scientific methods could gain the most traction. The supremacy of natural laws may be thought of as the uninterrupted uniformity of natural processes. If natural laws remain undisturbed by supernatural intervention, then the processes following those laws must remain uniform. And if there is no such interference, operations of nature in the past must have been the same as those observed today. In accordance with the deistic manner of thought, uninterrupted uniformity would extend back to the completion of creation, or as has been said, that "all things have continued as they were from the beginning of creation."[12] Uniformity of processes served as the foundation for the science of geology, which emerged at the end of the eighteenth century. Geology was the first science to attempt to reconstruct a natural history; the continuous processes observed in nature (erosion, sedimentation, etc.) were projected backward in time to explain past geologic change. The uniformity assumption (and it was an assumption) is what enabled science to extend the power of scientific methods into the past in order to explain the past. But it was an assumption brought to geologic observations, not obtained from them.

This point about an assumption brought to observations cannot be overstressed when one is trying to make sense out of the evolution-creation controversy. As much as science and the ideal of the scientific method implies the contrary, it is understood that the uniformity idea (otherwise known as uniformitarianism) was brought to the new science of geology to make sense of and explain the earth's rocks; it was not an idea obtained by observing rock formations. The deistic philosophy and theology of eighteenth-century Enlightenment lent validity to the assumption of uninterrupted uniformity, which then became the gateway for science to lay claim to natural history. It is likewise understood that ideas about catastrophic events causing the rock formations, such as the idea that the globe-wide flood described in the book of Genesis in the Bible caused them, are just as much brought to observations of the rock formations to explain them as is the uniformity idea. So how can the correct explanation be known? We will resume consideration of this question in chapter 3.

The march of science did not stop with uniformity since the creation. For all practical purposes, science has made the world safe from its creator. Science alone has come to be equated with objective knowledge because only science, it is thought, involves both logical reasoning and evidence. Historian of science Toby E. Huff remarks, "We in the West take the scientific point of view as the standard by which all others are to be judged."[13] Creation could not escape the category of religion, which many began to regard as a realm of subjective thought that did not need to be grounded on either logical reasoning or evidence. Science officially parted from philosophy in the mid-nineteenth century when the term *science* was proposed by William Whewell and widely accepted as a term of distinction among natural philosophers. Though whether science really had broken free of philosophical entanglements remains a relevant question.

WHAT A BIGGER PICTURE REVEALS

Did a culture given to a form of Christian theology have anything to do with the success of science? The question is appropriate because in the fourteenth century, both the Islamic Empire and China were more scientifically advanced than Christian Western Europe. The factors that led to the scientific revolution in Western Europe, while scientific progress collapsed in these other civilizations, are worthy of consideration. Historians have not arrived at a consensus opinion about the factors particular to Western Europe that set in motion the scientific revolution there as compared to elsewhere. Nevertheless, certain points may be noted with some confidence. Scientific advances in the Islamic Empire were first generated by contact with the writings of ancient Greece beginning in the eighth century. It was this contact that also inspired the beginnings of scientific advance (as natural philosophy) in Western Europe from the twelfth century. Whereas Islamic theology accorded Aristotle some superficial acceptance, the rationality of Aristotle was adopted by and integrated into Christian theology at a foundational level in the thirteenth century, most significantly through the writings of Thomas Aquinas (1225–74).[14] The union between Christian theology and Greek philosophy is generally considered the height of intellectual achievement during the late Middle Ages. So as rational inquiry into the natural world began to emerge three centuries later, it was considered, *in principle*, fully compatible with the reigning theology. Scientific pursuits were permitted as a peripheral activity in Islamic society for many centuries, but autonomy from the dictates of religion was never conceded. Islamic theology did not make peace with the scientific mind. Neither was there an acceptance of the ancient Greek philosophical idea that the natural world was ruled by a rational ordering of cause and effect.[15]

In the case of China, there was no direct contact with the writings of Greek philosophy. Though there was contact with the mathematical and scientific achievements of the Islamic Empire, that knowledge was not assimilated into Chinese use. Chinese science and mathematics developed independently. It is noteworthy that Arab astronomers were employed at the Chinese bureau of astronomy in Peking from the thirteenth century because the Chinese had not developed a system of trigonometry essential for making calculations in astronomy. But as with Islam, a rational approach to understanding the natural world through scientific inquiry was never granted status or autonomy within Chinese culture. The natural world was viewed more as something with which both individuals and the social order had to be in balance and harmony, not as an object of study. Chinese society exalted social harmony and conformance rather than individual initiative and achievement. Scientific discoveries were not valued enough to be preserved, exploited, and built upon.

Given this wider historical perspective encompassing other cultures, the confrontation between Galileo and the Catholic authorities should be viewed as a ripple in an otherwise compatible coexistence and interaction between natural philosophy and Christian theology before, during, and after the Renaissance. The rationality of Greek philosophy was embraced by Christian theology; it was not generally seen as a threat to that theology.

Sustained acceptance of rational inquiry into the natural world did not happen where different theological or sociocultural contexts did not permit it. Suppression of ideas contrary to a reigning ideology is characteristic of culture. The predispositions that define a culture and justify its ideologies also determine what thoughts are palatable. The words *cult, cultivate* and *culture* have a common root. Though their usages are different, when applied to the realm of ideas, the meanings are not far apart. What history is showing is that a culture reared on Christian theology did not inhibit the reasoning mind as other cultures had, but in fact encouraged it. What science owes to belief in a rational world will be discussed further in chapters 8, 9, and 10.

SUMMARY

The following five points summarize this historical overview of early science and the scientific revolution: (1) Acceptance by Christian theology of the rational approach of Greek philosophy to understand the natural world produced natural philosophy. (2) Natural philosophy restricted to immediate causes, while excluding purpose and design, became the mechanical philosophy. (3) Both empirical verification and mathematical characterization of natural cause and effect were joined to the mechanical philosophy. (4) The resulting mechanical philosophy, now essentially science as practiced today, led to the philosophy of deism. And (5) the deistic understanding of the created order and the corollary uniformity principle enabled science to lay claim to natural history.

The history of ideas is intriguing. Some ideas were joined together; others were drawn apart. Certain trends represented real progress. With other trends, that there was progress might be disputed. Yet the history of early science clearly shows that the enterprise of science was predicated on a natural world whose observed consistency and ordering is attributed to a rationality understandable by the mind. The same remains true today. Huff further remarks how it is seen that science is "both sponsored by and motivated by the idea that the natural world is a rational and ordered universe and that man is a rational creature who is able to understand and accurately describe that universe."[16] The natural world can be figured out! Observation and reason became the twin pillars of scientific knowledge and the scientific method. Mathematically reasoned and empirically verified, this particular formulation of inquiry has been extremely successful in gaining insight into the natural world. But in the confidence of its success, has science overstepped the bounds of empirical verifiability that it set for itself?

Chapter 2:
Genesis

*All teaching, preaching, and instruction in the Word is predicated
on the realization that it says one thing and not another.*
—Anonymous

Of course, science is not all there is at the center of the controversy over the origin of the universe, earth, and all living things. So is Genesis 1. It is worthwhile to take a look at its history also, or rather the history of its interpretation.

It is possible to anticipate or desire a certain outcome or conclusion when one is interpreting a particular passage of the Bible, but that should not lead to a misunderstanding about what is being communicated. As an example, one might ask, "How should the first chapter of the book of Genesis be interpreted in light of the discoveries of science that suggest the earth is approximately 4.6 billion years old and that evolution has occurred?" But this is not a question we should be asking. In order to arrive at what Genesis 1 truly says, preconceived ideas and external influences about what it should say must be put aside. Scientific conclusions and theories need to be put aside, at least temporarily. Otherwise, it may be impossible to determine what the author of Genesis is communicating. Trying to read a position into a passage of scripture or adjusting the meaning of the words to accommodate a preconceived idea is dishonesty. Stressing certain passages while ignoring others in order to obtain a "correct" answer is also dishonest. We should start with one presumption:

The author did not intend what he was communicating to be a mystery. Clearly, the validity of this presumption rests on whether a coherent and understandable meaning is forthcoming, but literature is never approached with any other expectation.

The intent of this chapter is to objectively answer two questions about the interpretation of Genesis on certain matters where there has been disagreement:

1. What do the scriptures say about the age of the earth and universe?
2. Is theistic evolution an interpretation compatible with the scriptures?

INTERPRETATION ON THE QUESTION OF AGE

The means to arrive at the age of creation is straightforward and is based primarily on the genealogy records in Genesis 5 and 11. The numbers of years from birth to birth of the various patriarchs in these chapters enable one to add up the number of years from Abraham back to Adam.[1] Because the approximate time of Abraham can be figured from elsewhere in the Bible, the total number of years counting backward to Abraham and then to Adam gives the approximate age of creation. (If the six days described in Genesis 1 are days as defined by one rotation of the earth, they would not increase the total number of years.) Old Testament scholarship recognizes some uncertainty on the recorded genealogies because of the possibility that some generations may be missing. There is not complete agreement among scholars on an exact age, even assuming no missing generations. However, the range of uncertainty is not great; an age of slightly more than six thousand years is generally thought to be a correct calculation from the genealogy records. If there are missing generations, the age of creation may be somewhat greater than the calculations indicate. The earliest records of civilization (dating about 2900 BC and 3200 BC from Sumeria and ancient Egypt, respectively) are generally consistent with the estimated date of the flood described in Genesis 6 through 8, which, assuming no missing generations in Genesis 11, would have occurred at about 2500 BC. Any historical records existing prior to the flood would presumably have been destroyed in it.

These calculations are not difficult, yet there are few indications of attempts prior to the sixteenth century.[2] The age question was perhaps not a matter of serious contemplation until naturalists sought to answer it. In the sixteenth and seventeenth centuries, the Bible was the logical and appropriate source to turn to for an answer. In 1583, Joseph Scaliger calculated an age for the creation of approximately 5,500 years (at that time), having assumed no missing generations.[3] (Note that the calendar was revised to its present modern form the previous year, 1582, by Pope Gregory XIII.) The most referenced date of creation is the 1620 calculation by Bishop James Ussher of 4004 BC. In this, we have as a historical reference point, an interpretation, what may be called the traditional or literal interpretation, free from the influence of scientific conclusions because there was no such external bias in the late sixteenth through the mid-seventeenth century toward any particular answer. It was not until the end of the eighteenth century about two hundred years later, as the science of geology was birthed, that naturalists began to infer an earth several million or more years old.

Is the traditional interpretation a sound interpretation? To answer this question, we should consider the available alternatives and the challenges they present to the traditional interpretation. Interpretations that attempted to extend the beginning of creation earlier in time by several million years appeared after naturalists began to interpret geologic observations with reference to the principle of uniformity in the early nineteenth century. They theorized that geologic processes observed in the present are the same as what occurred in the past. Accordingly, rock layers must have been built up in the past slowly over long periods of time. Applied to geology, this use of the principle of uniformity is called uniformitarianism. Some of the interpretive issues that arose were with respect to how much time passed before the first day began and what exactly is meant by the word *day*. It might be said that some previously unrecognized gaps and rather long days began to pop out of the first chapter of Genesis as the traditional interpretation was revisited. We should take note that these alternate interpretations, such as the gap, day-age, progressive creation, and divine fiat theories, did not appear until after naturalists began arriving at uniformitarian conclusions about the age of the earth. In fact, they came in direct response to uniformitarian ages as an attempt to harmonize scripture with science. Here are the known alternate interpretations:

- Gap theory: an undescribed creation and destruction between Genesis 1:1 and 1:2.
- Day-age theory: each of the six "days" spans an indefinite length of time or age.

- Progressive creation theory: multiple creations of progressively advanced organisms spaced out over geologic time as correlated to fossil-bearing rocks.
- Divine fiat theory: commands of God in Genesis 1 spanned six days but were not fulfilled until much later.
- Days of revelation to Moses theory: God explained the creation to Moses in six or seven days.[4]

The fact of these theories' appearances in conjunction with uniformitarian geology is sufficient cause for them to be viewed with suspicion. But for the moment, let us put aside that suspicion. The day-age theory at least is still seriously considered and widely believed. Is the word translated *day* from the Hebrew language a day or something else? Granted, this word in the original language is general enough to mean something other than a day.[5] In such a case, the context determines the meaning. In calling attention to the context, we should consider the meaning of Genesis 1:2 and the location from which God spoke. The location from which God spoke the creation into existence becomes important for establishing the intended meaning of the word *day* as used in Genesis 1. Genesis 1:2 (RSV) says, "The earth was without form and void, and darkness was upon the face of the deep; and the Spirit of God was moving over the face of the waters." Does this mean that the Spirit of God was in heaven, out in space, on earth, or what?

Certain other passages, specifically Psalm 104:1–9, 2 Peter 3:5, and Job 38:4–11, help the reader in arriving at a reliable comprehension of Genesis 1:2. In Psalm 104:6, parallel usage of the terms *the deep* and *the waters* conveys a picture of oceans covering the earth at the beginning of creation. In context, this does not appear to be a reference to the flood of Genesis 6 through 8, though a subsequent verse, Psalm 104:9, might refer to bounds for "the waters" set in place after that flood. 2 Peter 3:5 confirms that water covered the earth at the beginning of creation. The clarification introduced by these two passages is consistent with Genesis 1:6–10. In Genesis 1:6–10, the waters above were separated from the waters below (later called seas), resulting in the firmament (called heaven) on the second day. On the third day, the sea waters were gathered in one place so that dry land would appear. Note that Psalm 104:3 uses the terms *clouds* and *waters* in a similar parallel fashion as verse 6 uses *the deep* and *the waters* clarifying that the waters above the firmament were clouds.[6] From Job 38:9, it would appear that the darkness noted in Genesis 1:2 was the result of a thick cloud covering. These other passages help to establish that "moving over the face of the waters" means at the surface of the earth—that is, over the seawaters and under the cloud waters.

While there is no interpretive aid from other Bible passages on the other aspects of Genesis 1:2 (i.e., the earth being without form and void), interpretation may align with what is already known. The earth's lack of form might mean that there were no land formations, this being the case if everything was covered by water. This could be taken as one possible interpretation, yet it is at least consistent with the verifiable point of a total water covering. That the earth was void might mean that there was no life yet. Again, this is a possible interpretation but one that is consistent with what follows in Genesis 1.

Having identified that the Spirit of God was "moving" at or near the surface of the earth, the evenings and mornings associated with each of the six creation days should lead one to consider that *day* means one rotation of the earth. The earth being spherical, both day and night are present somewhere all the time.[7] A single point at the surface of the earth is the only location from which evening and morning can be experienced. Evening and morning cannot be experienced from any other location. There are no days and nights or evenings and mornings out in space. A location at or near the surface of the earth is also the only one that would have been comprehensible and familiar to Moses and the Israelites.

Could the evenings and mornings refer metaphorically to endings and beginnings of longer time periods, such as in "the dawning of a new age" or "the sun had set on the empire?" Certainly, but the most straightforward interpretation here is the literal one. If it was not for the location established by Genesis 1:2 and the above Old and New Testament verses, evening and morning would be more open to metaphorical interpretation.

Discussion on the meaning of the word *day* would be incomplete without considering what is behind the fourth commandment in Exodus 20:8–11. Exodus 20:11 states that the six creation days and the seventh rest day, narrated through Genesis 1 and continuing through the first few verses of Genesis 2, set the pattern for the seven day week. The week is to consist of six days of work and then the Sabbath day. The six creation days and the seventh rest day define the week; any other meaning of *day* in Genesis 1 nullifies the justification for the Sabbath day commandment given in Exodus 20:11. In fact, on this point, we do not need to rely on our own interpretation of *day*; we have the interpretation of Moses and the Israelites preserved through the ages. The seven-day week, so integral to the modern calendar, is a continuing testimony to the authority of that interpretation of Genesis 1. Any other sense of the word *day* in this passage overall would mean that the Jews were and still are seriously mistaken regarding the intent of the fourth commandment.[8] Psalm 99:6–7 confirms that Moses was right. The Exodus 20 passage, therefore, excludes the possibility of the word *day* meaning anything other than one rotation of the earth. A long time period or age is not what *day* means.[9]

The possibility of an indefinite time period between Genesis 1:1 and 1:2, "the gap," as it is called, is also disqualified by Exodus 20:11 which states that heaven and earth were included in the six days of creation. What this verse is saying is that the period of time during which the "earth was without form and void, and darkness was upon the face of the deep," though perhaps preceding the first day, was short enough to not require more time or more days to be added to the six creation days.

In Genesis 1, a pattern is established wherein each day begins with "And God said—" and concludes with "And there was evening and there was morning—" Within each day, God says, "Let there be—" That which was created appears or was so, and He "saw that it was good." The possibility that the commands spoken by God were not fulfilled until sometime after the day they were uttered, according to the divine fiat theory, is excluded because "it was so" and He "saw that it was good" are all within the boundaries of what the text demarks as each day.

The early chapters of Genesis have a simplicity of style and terminology. All the things described would have been familiar to Moses and the Israelites: light, water, land, plants, seeds, sun, moon, stars, birds, fish, and livestock. No gamma rays, bacteria, spiral galaxies, or strange deep-sea creatures are mentioned, although they exist. The most straightforward and defendable position is that what was written by Moses would have been readily understandable to him and his audience as it was given.[10] This position merits priority in interpretation because it is the one most consistent with our one presumption noted in the introductory paragraph above. There is no indication from the text that would imply that the meaning was hidden or that it would not become apparent until after other information was supplied. Neither is there any indication that this text would have required scientific knowledge to comprehend it.

The answer to the age question is that the scriptures permit an estimate on the age of creation to be made. Calculations from the available genealogical data indicate that the earth and heavens are at least six thousand years old. This age could be extended to account for possible missing generations in the Genesis 5 and 11 genealogies. How much of an extension might be justified is anyone's guess, but more than an additional few thousand years would seem an unreasonable appeal to missing generations.

ON THE QUESTION OF THEISTIC EVOLUTION

What about theistic[11] evolution? Theistic evolution is the proposition that God created by means of evolution. It was devised in the 1860s as a compromise between divine creation and Darwin's 1859 evolutionary hypothesis. It was not so much a new interpretation of Genesis as it was an initiative to bridge natural theology (or the argument for design or divine purpose in nature) with the evolutionary ideas of randomly arising variations and natural selection in an intellectually satisfying fashion. Because of the earlier challenges by geology and associated interpretive compromises, a nonliteral view of the Genesis timescale had already become the established view on the age of the earth before the 1860s. So there was no great leap involved in linking theology and science in this manner.

However, the idea of evolution does not arise from Genesis 1. Prior to evolution becoming a scientific theory, no one ever interpreted this text as suggesting, even in the slightest degree, the idea of a continual process of incremental development over long periods of time. With its daily initiations and completions of each of the types of things brought into existence, the text runs contrary to the idea of incremental development over long periods of time. The narrative of Genesis 1 tells us that the evolutionary aspect of theistic evolution has no place, if only because of the manner in which *day* must be interpreted.

On the other hand, top evolutionary scientists tell us that the theistic aspect of theistic evolution is unnecessary, that it is extraneous. They contend that evolutionary science stands securely on natural causes alone and does not need a creator, or they at least claim that science can say nothing about the existence of God or a creator. We are left, therefore, with no positive grounds from either camp for believing in theistic evolution. Neither scriptural grounds for the evolutionary aspect nor scientific grounds for the theistic aspect exist. This attempt at resolution can survive only as a compromise position whose appeal is an end to the controversy.

Theistic evolution is a popular view but not an interpretation that is compatible with the scriptures. The theory of progressive creation as a series of creations spaced out at intervals so as to correlate with the uniformitarian explanation of the fossil containing rock strata falls by the same argument.

On Figurative or Subjective Interpretation

I n an attempt to dodge the above two conclusions, an alternate interpretive approach to the first few chapters of Genesis is to presume their meaning is figurative or that they constitute an allegory not to be interpreted literally or considered as historical. What merit does this interpretive approach offer?

Figurative language is used to describe intangible concepts in a tangible way; it helps to communicate ideas and abstractions. Assuming for the moment that these chapters constitute an allegory, we need to ask what the real message behind the allegory is. What idea or abstraction is being expounded? It is true that certain abstract ideas about humanity are explained in the first three chapters—freedom of will and the nature of evil, for example. But that does not mean that a literal and historical interpretation must be discarded and a figurative basis substituted. It might be said that the idea of beginning itself is the idea being expounded. But that can work as well in a literal and historical mode as in a figurative mode; the point is inconclusive.

It is noteworthy that there is no transition of style in the narrative from the creation narrative to events chronicled in later chapters, the historical nature of which is not questioned. In other words, there is no point that could be identified as an ending to allegory and beginning to history. In fact, there are indications elsewhere in the Bible that the creation narrative is not figurative or allegorical but historical:

1. In Matthew 19:4–6, Jesus refers to Genesis 1:27 and 2:24 as historical (nonfigurative) in making a point about marriage.
2. Luke traces Jesus's physical ancestry back to Adam in Luke 3:23–38.
3. Paul refers to Adam in 1 Corinthians 15:21–2 and 15:45–9 as a historical person as he also refers to Christ.
4. Exodus 20:8–11 refers to the six days of creation in a way that signifies a literal meaning by their correlation to the seven-day week.

There is then not only an absence of indications that the authorship intended the early chapters of Genesis to be taken allegorically but indications that they are not so intended. Any conclusion as to the intended meaning based on considerations external to the scriptures should be regarded as suspect. You try to figure out what an author is saying by what he wrote and not by what someone else says he meant to write. Therefore, these chapters should be regarded not as an allegory but as a record of historical events.[12]

Finally, there is a view that implies that all interpretations of scripture are subjective and that no objectively correct interpretation is possible. This view may seldom be stated explicitly but rather suggested in such a manner as to bypass rational judgment: "Many viable interpretations are available. How can any one of them be more correct than another?" To imply that objectivity is unattainable is in opposition to the one presumption noted at the beginning of this chapter, that authors do not intend what they are communicating to be a mystery and that this is never the manner in which literature is approached. Chiefly, the preceding arguments toward a single interpretation should demonstrate that objective interpretation is possible.

CONCLUDING REMARKS

The preceding arguments may be of no relevance at all (or of relevance only in accounting for certain historical developments) if the book of Genesis and the Bible are, as some claim, make-believe. It is true that many accept the authority of Genesis and the Bible without knowing why they should accept it. That belief is just something handed down to them. Moreover, independent verification of the historical accuracy of the book of Genesis, particularly the first several chapters, is not possible. Nevertheless, there are grounds for accepting the Bible as authoritative and, as integral with it, the book of Genesis as well. These grounds are introduced in chapters 11 and 12 of this book.

The conclusions reached in this chapter include the following: (1) The scriptures are not silent on the age of creation and that the earth and universe can be determined to be at least six thousand years old, though probably not much more. And (2) theistic evolution has no scriptural grounds and appears to exist only as a compromise position whose appeal is an end to the controversy. These conclusions on what Genesis 1 says are not inerrant. While the scriptures themselves are inerrant, interpretations and conclusions are not. Therefore, the relevant scriptural passages have been identified, and the reasoning from them has been presented to show how the conclusions were reached. Other interpretations of Genesis 1 could be considered, though at this point, no others are known to have been proposed.

Having validated the traditional interpretation of Genesis 1, we now turn to consider the ramifications of what the scriptures are saying and the conflicts with scientific theories. Discoveries and developments in the sciences of geology, biology, and astronomy have appeared to seriously contradict the above scripturally derived understanding of the origin of the natural world, the earth, and the universe. Until the early nineteenth century, naturalists were generally committed to the authority of the Bible, that the earth was a few thousand years old, and that it was created. There was a reluctance to devise theories that might have caused a negative reaction from church authorities or the public. But in view of the situation existing between science and the Bible today, a conflict of understanding somehow came about. The next several chapters will look into the histories of these sciences as they pertain to the developments resulting in that conflicting understanding.

Chapter 3:
Geology

*The result, therefore, of our present enquiry is that we find
no vestige of a beginning—no prospect of an end.*

—James Hutton,
the father of modern geology

I n chapter 1, we saw how a principle was introduced in association
with a scientifically inspired eighteenth-century philosophy. That
principle is the idea that all that has happened in nature since the creation
needs only to be explained by operations of the laws of nature. It is that
principle that eventually enabled scientists to consider natural history and
devise explanatory theories on what happened in the unobserved past.
The principle is also described as uninterrupted uniformity of natural
processes. It was first applied by early geologists more than two centuries
ago as they began to study the earth's rock formations and seek scientific
explanations for them. Uninterrupted uniformity came into conflict with
the general assent that the miraculous worldwide flood of the Genesis
record was somehow responsible for much of the earth's geology. Among
the sciences, geology was the first to challenge the authority of the Genesis
creation record. Today, the assumption of uninterrupted uniformity has
eclipsed even the idea that there was a creation.

Beginning with Galileo's discoveries in the early seventeenth century,
the importance of observation to scientific inquiry was no longer overlooked
as it had been. After Galileo and Bacon, naturalists sought to validate their

theories about nature by observable fact. But this transformation was often slow, and geology was not one of the leading fields in the progress of scientific knowledge. Natural philosophers were devising explanations on what natural processes and events might have been responsible for mountain ranges and rock layers, but until the late eighteenth century, explanations were speculative with minimal reliance on detailed observation. This began to change with the industrial revolution. The period between 1790 and 1830 is sometimes referred to as the heroic age of geology as this area of study moved to the forefront of scientific advance.[1]

Naturalists now began to study rock layers or strata as they appeared in different exposed outcroppings, to observe the order of their stacking, and to map their geographic extent. As a first instance of this at about the year 1800, William Smith (1769–1839), while surveying routes for canals in England, used the presence of fossils imbedded in rock strata to identify and correlate rock layers from one location to another. Wherever he found the same type of fossils in particular strata located in different areas, he concluded that they must be the same stratum. New geologic observations generated further academic interest in the causes of rock formations and their age. Beginning late in the eighteenth century and continuing into the nineteenth century, new theories were proposed in an attempt to explain the origin of rock layers and mountain ranges.

We should note that it was still widely believed that truth discerned from nature somehow had to agree with the truth of scripture. The agreement of new geologic theories with Genesis was a desired outcome, though as it rather turned out, it was Genesis that was made to agree with the new theories. As we saw in chapter 2, several new interpretive theories on Genesis were devised at this time in order to reconcile the creation account with the latest geologic theories.

TWO CONTRASTING THEORIES

Two contrasting geological views competed during the heroic age of geology: catastrophism and uniformitarianism. The catastrophist theory said that sudden and extreme forces and events were the chief factors involved in the formation of rock layers and mountain ranges. Catastrophic events, specifically the Genesis flood as described in Genesis 6, 7, and 8, but also other catastrophic floods as well as earthquakes and volcanoes were seen as necessary to explain rock formations. At first, the Genesis flood was assumed to be responsible for everything. But as scientific findings on the multiplicity of rock strata accumulated, so did the complexities of catastrophism. Each newly found geologic formation needed its own separate explanation. The numerous distinct rock layers seemed highly unlikely to have been caused by one flood; in some locations, numerous layers of limestone, sandstone, and shale are interspersed in their order of stacking. Both catastrophists and uniformitarians became convinced that the great number of distinguishable sedimentary layers of great thickness resulted from separate events or time periods. The wide variety of fossils of extinct organisms, many of them segregated according to type within the different strata, also seemed to indicate that the layers were formed by separate events widely spaced in time.[2] As explanations moved away from a one-flood model, the amount of time involved to accommodate a series of catastrophic events could no longer be reconciled with a six-thousand-year-old earth. A literal interpretation of the Genesis six-day timescale was abandoned, but miraculous intervention remained acceptable. Underlying the catastrophist theory was the position that geologic processes observed in the present day were entirely inadequate to explain the geology.

The contrasting approach and ultimately the more successful of the two schools of geologic thought was the uniformitarian view. This theory maintained that the processes of erosion and deposition in the past were essentially the same as observed today and that these processes constitute a sufficient explanation for the observed rock layers. Although the layers are numerous and often massive, the rate of deposition observed today (chiefly on seafloors, continental shelves, lake bottoms, etc., or infrequently from volcanic eruptions) is the rate at which they were thought to have been formed. The uniformitarian theory was first expounded by James Hutton (1726–1797).[3] Hutton's characterization of the uniformity principle that "the present is the key to the past" was the theme of his 1785 *Theory of the Earth*. The idea was further expounded and supported with the availability of more empirical data by Charles Lyell (1797–1875) in the 1830s. Lyell's *Principles of Geology* was published in three volumes between 1830 and 1833 and was thereafter a major influence and the reason for the predominance of uniformitarianism over catastrophism. Lyell's *Principles of Geology* has as its subtitle *Being an Attempt to Explain the Former Changes of the Earth's Surface by Reference to Causes Now in Operation*. This subtitle is about as succinct a definition of uniformitarian geology as you will find.

Far more than catastrophism, uniformitarianism required substantial lengths of time. This is because it was necessary to assume the same rate of erosion and deposition as observed today to explain the immense thicknesses and quantity of sedimentary rock layers. The idea that rock layers were built up over long periods of time on seafloors, etc., was a foregone conclusion given an uniformitarian premise. At the bottom of seas is where deposition is observed today, and normal buildup of sand, silt, shells, etc., as also observed today, is very slow. The appearance of uniform and distinct layers suggested calm conditions, and long time periods would obviously have been necessary to form numerous and massive rock layers under such conditions. One often encounters references to "warm, shallow seas" involved in the formation of the massive rock layers of the Colorado Plateau, for example. On this premise, hundreds of millions of years were necessary to explain the formation of the observed rock layers. Thus, uniformitarianism, more so than catastrophism, constituted a serious breach of the widely acknowledged authority of scripture up to this time. Uniformitarian geology was also laying a foundation for further breaches of scriptural authority in biology that were to arise a few decades later. As a result of the predominance of uniformitarian geology in the 1830s, the idea of vast geologic ages was well entrenched in scientific circles by the time Darwin published *Origin of Species* in 1859.

Conceptual predispositions often rule the interpretation of empirical findings, and this also was true two hundred years ago. At that time, the uniformitarian school insisted on rigid adherence to the ruling principle that only currently observed geologic processes were permitted. Such findings as geological unconformities—that is, locations where the actual stacking of rock layers was contrary to the usually observed stacking or where layers were missing—were not easily explained. Yet such additional relevant facts could be overlooked in deference to a unifying principle. The desire to hang on to a unifying explanation, especially after it has been seen to be workable in explaining a majority of the data, is considerable. Modern geology includes both uniform processes and catastrophic events, but uniformitarianism had a definite appeal among nineteenth-century naturalists because of its link with observation. Once the uniformitarian premise was established, catastrophism's favored causes were allowed to trickle in as supplemental factors, and all major geologic discoveries were eventually explained in a satisfying manner.

THE GEOLOGIC COLUMN

A most significant product of the uniformitarian premise concerns the ordering of rock layers and is known as the geologic column. The geologic column is a construct of the earth's rock layers of sedimentary and volcanic origin as observed from bottom to top. It was developed piecemeal throughout the nineteenth century as geologic observations in Europe and later from around the globe were collected and correlated. Geochronology is the branch of geology dealing with the geologic column and geologic time. An all-important assumption is that the stacking from bottom to top corresponds to oldest to youngest, that bottom layers were deposited first and then each layer above in succession. The complete column occurs nowhere, but stacks of rock layers of various sorts do appear all around the earth.

A stack of layers in a particular locality is known as a type section. Type sections can be correlated with each other whenever layers can be identified as common to two or more locations. Identification and correlation of common layers within close geographic vicinity can usually be determined by the type of rock (sandstone, limestone, shale, etc.), by characteristics, such as color, grain size, etc., and also by fossil content if there are fossils contained within the layer. However, the validity of correlations over geographically remote areas depends on how geographically extensive the various layers are. Because no formations exist clear around the globe or even over whole continents and because formations are generally confined to smaller geographic regions, the use of fossils to make the correlations becomes necessary. If there is similar fossil content, any two formations being compared may be assumed to have been formed within the same geological time period. If the fossil content is different, even if other characteristics are similar, the rocks are assumed to have been formed during different time periods.

Fossils then are the only means of correlation on a global or even a continental scale.[4] The specific types of fossils that make the correlations possible are called index fossils. What makes for a good index fossil is one with very wide, preferably global geographic distribution, but at the same time, one confined to one or only a few layers in any given type section. An essential premise for constructing the geologic column is the validity of progressive evolution or faunal succession; global correlation and stacking based on fossil content depends on knowing the positions of index fossils in an evolutionary time sequence.

RADIO-ISOTOPE DATING

The order of stacking determines relative ages. Given any two adjacent layers, the one on the bottom is assumed to be older than the one on top, but that is all that can be determined. Until the twentieth century, no means of estimating the absolute ages of rock layers was available, so their absolute ages remained unknown. The discovery of radioactivity in uranium by Henry Becquerel (1852–1908) in 1896 and the formulation of atomic theory by Lord Ernest Rutherford (1871–1937) in 1905 led to a first attempt to estimate geologic ages based on radioactivity measurements of rock samples in 1907. Through the first half of the twentieth century, the use of decay rates of certain radioactive isotopes as a means of obtaining estimates on the absolute ages of rocks was further investigated and refined by Arthur Holmes (1890–1965).

An isotope is an atom of a certain element for which it is possible to have different atomic weights. The difference in weight is due to having extra neutrons in the nucleus. A radioactive isotope is an atom having extra neutrons that may be lost through a spontaneous process of decay. In a sufficiently large sample of these atoms, the decay rates become predictable with high accuracy and precision. Most radioactive isotopes have fast decay rates, but a few have rates measured in hundreds of millions and billions of years. The rates of decay are assumed to have been consistent over time because the atomic forces are so strong in comparison to any environmental factors, even tremendous heat and pressure. Experiments have shown that heat, pressure, and radiation do not appear to affect the decay rates of isotopes. The following isotopes are specifically used in geologic dating:

- Uranium 235 decays to lead 207: half-life = 704–713 million years
- Potassium 40 decays to argon 40: half-life = 1.25 billion years
- Uranium 238 decays to lead 206: half-life = 4.46–4.51 billion years
- Thorium 232 decays to lead 208: half-life = 14.0–14.1 billion years
- Rubidium 87 decays to strontium 87: half-life = 48.8 billion years

Radioactive decay is exponential, and the half-life is the amount of time it takes for exactly one half of the remaining isotope atoms in a sample to decay from any point in time. Once a rock formation solidifies from a molten state, the isotopes are presumably locked up, and the radioactive decay process from then on serves as a type of clock. This property allows estimates of a rock's age, given certain assumptions: (1) radioactive decay rates have remained unchanged over time, (2) concentrations of isotopes at the time the rock was formed can be reliably inferred, and (3) no significant amounts of the isotope element itself or the element into which it decays have leaked out or otherwise found their way into the rock sample being measured since its formation.

These assumptions together with measurements of the concentrations of isotope elements and their decay product elements enable the age of a rock sample to be calculated. The first assumption is a deduction from the uniformity principle, but as far as present experimentation on decay rates can verify, the assumption appears to be sound. Assumptions on the relative concentrations of the isotopes present upon initial rock formation and the possibility of loss (or addition) of one or the other isotopes since the time of rock formation are not as easily validated. The reliability of assumptions on initial concentration and possible leakage depends on study of the formation from which a sample is obtained, how it was likely formed, and what it may have been subjected to since. Geologists have gained extensive experience with the radio-isotope methodology over the past century and are confident that their assumptions on these two points are also sound.

The isotope elements listed above are rare, and so the concentrations in rock samples to be dated are usually very small. But it is still possible to obtain accurate measurements of small isotope concentrations as long as the sample of rock being tested is big enough. The larger the rock sample to be tested, the more reliable the measurement of the isotope content. A further difficulty with radio-isotope-dating methods is that igneous (volcanic) rocks are the only ones that can be dated using radioactivity. Igneous rocks contain no fossils. The ages of sedimentary layers, the ones containing the fossils, can be estimated only if there are interspersed volcanic layers or if there are igneous intrusions into the sedimentary layers. Sedimentary rocks cannot be dated directly because the material comprising them was either eroded from other preexisting formations or, as in the case of limestone, is an accumulation of marine animal remains. In order to improve the reliability of age estimates, two or more of the listed isotopes are used to date a given sample where possible. However, it is often the case that variations in isotope concentrations from samples taken from the same rock formation give scattered age results and that the different isotopes can also give conflicting estimates.

The usefulness of these isotope-dating techniques is limited to measuring ages that are at least a significant fraction of the half-life. Any attempt to measure rocks of comparatively more recent origin ends up producing meaningless results. It would be as if, for example, a calendar were used to time a hundred-meter dash. The estimate on the age of the earth is based on a number of different approaches involving measurements of isotope concentrations among which are the following:

1. Several long-lived isotopes besides those listed above, though with somewhat shorter half-lives, are completely absent in the earth and sun. Yet these isotopes are observed in other stars and interstellar matter by means of spectroscopy. After about fifteen half-lives, practically all unstable isotope atoms of any element will have decayed out. Assuming that these isotopes previously existed in the earth and sun, an age of at least one billion years is indicated.[5]

2. The uranium 235 isotope has a concentration of 0.7 percent observed in all uranium today. When uranium was first formed, the concentration of uranium 235 is thought to have been as much as but no more than 50 percent. This puts an upper limit on the age of the earth at six billion years.[6]

3. The assumed buildup of each of the three lead isotopes listed above in the earth's crust through radioactive decay yields comparable ages and puts a maximum age at 5.5 billion years. Rocks not containing uranium or thorium are used as measures of the original lead isotope concentrations in the rocks being measured.[7]

4. Relative concentrations of uranium and lead isotopes in meteorites are applied in the following manner: Iron meteorites contain very little uranium, so it is assumed that the lead contained in them is original and not the result of radioactive decay of uranium. Stony meteorites contain enough uranium to assume that much of the lead in them has resulted from radioactive decay. If the amount of lead in the iron meteorites is subtracted from that found in the stony meteorites, the difference is assumed to have been the result of radioactive decay of uranium in the stony meteorites. This yields an

estimate of 4.5 billion years, assuming that meteorites were formed at the same time as the earth.[8]

It is important to be aware that carbon dating is not the same as these forms of radio-isotope dating. Carbon-14 has a half-life of 5,715 to 5,730 years and can only be used to date organic materials. It is thus notably different than these other methods that apply to inorganic rock. For example, C-14 dating has been used to help date the scroll material of the Dead Sea Scrolls to about the first century BC. A concentration of C-14 in the atmosphere is continuously maintained by the encounter of cosmic radiation with nitrogen atoms, and this concentration level is assumed to have been consistent over time. Plants absorb C-14 in the process of exchanging carbon dioxide with the environment. Animals pick up C-14 when consuming plant material or through the food chain. Whenever any living plant material or animal dies, carbon is no longer exchanged with the environment, and the amount of C-14 is locked up. The C-14 clock then begins ticking. C-14 dating has an upper limit of validity of approximately fifty thousand years. After that amount of time, there would not be enough C-14 remaining to make reliable age estimates.

COMMENTARIES

The geologic column and the radio-isotope-dating methodology both merit some comment. The assumption behind the geologic column was mentioned above, that index fossils are the means of correlating rock layers over wide geographic areas. The stacking of the column is thus premised on an ordering of the index fossils, and this ordering is dependent on what biologists have determined to be the relative complexities of the index fossils or their evolutionary sequence. This means that the ordering assumes evolution from simple to complex index fossils. The simple fossils are assumed to be older than the complex, and if older, then the strata in which they are found are placed lower in the column. Likewise, the more complex are placed higher in the column. However, if the premise of evolution from simple to complex is withdrawn, this means of stacking the column beyond type sections disappears as well. If there is to be any worldwide stacking at all, the evolutionary premise seems to be required. Biological evolution is itself dependent on certain assumptions and should not be taken for granted when other assumptions could yield a substantially different conclusion in geochronology. If the earth is only as old as the book of Genesis implies and evolution is false, the geologic column is of little use. In that case, the only ordering of fossils that is relevant and that would need to be accounted for is whatever ordering is found in the type sections.

The estimate of the age of the earth is built on a complex of measurements, assumptions, and inferences involving radio-isotope-dating methodology, of which the above four points are a brief summary.[9] Radio-isotope dating was developed as a means of finding absolute ages of rock formations, but the conclusions reached by the methodology followed (by about a century) the view that the earth was hundreds of millions of years old. It is not inconceivable that measurement data, assumptions, and inferences fitting well with that view have been considered favorably. Nonconforming measurements could be explained away or discarded. While the use of radioactive isotopes for dating is straightforward in theory, the practicalities involved in applying the technique are not. Evaluating past measurements, assumptions, and inferences is probably beyond the capacity of anyone who is not a practiced geophysicist.

Weighing against the opposing case is the fact that no theory has been proposed that explains why the radio-isotope measurements are what they are if the earth is not nearly so old. Recent scientific research has found indications that radioactive decay rates may not have been constant in the past. One such indication is the presence of helium gas in zircon crystals. Helium is a by-product of uranium decay and one that easily leaks out. It is found in zircon crystals at levels too high to be accounted for by the normal decay rate of the uranium found in them.[10] However, until a coherent theory on past decay rates is established, it is acknowledged that radio-isotope conclusions contradict the book of Genesis's timescale.

A question may arise at this point as to whether or not, given the development of radio-isotope methodology, index fossils are still a necessary underpinning of the geologic column. Could not the geologic column be ordered exclusively by radio-isotope-derived dates? The answer is that index fossils remain necessary because of limitations in radio-isotope methodology, among which is the inability to date sedimentary layers directly. The methodology is not able to determine the order of the geologic column entirely on its own.

OTHER DATING APPROACHES

In addition to radio-isotope dating, many other natural process rates that rely on uniformitarian assumptions are used to help estimate the age of the earth. When considered independently, some of these processes tend to support a ten-thousand-year or less age. Examples include comet life spans, buildup of sediments on the ocean floor, and dissolved minerals in seawater. Other processes have been identified that seem to have run much longer than ten thousand years. Examples include continental drift, polar ice cap deposits, and coral reef buildup. These are only a few of the many dozens of time-tracking processes offered. Yet for most of these processes, counterarguments are advanced that are intended to discredit their validity for dating purposes. Do the arguments and counterarguments collectively make one or the other of the age of the earth positions more vulnerable? For a nonscientist to sort through the data, critically evaluating the validity of the uniformitarian assumptions advanced in support of the two positions would seem to be an overwhelming undertaking.

An example of one of these other natural process rates is that of coral reef formation. Coral reefs consist of skeletons of tiny organisms. The skeletons are built on top of each other in solid masses through succeeding generations. The fastest accumulation of coral observed is about a foot every thirty years. The Eniwetok Atoll in the Pacific is a reef 4,610 feet thick, which implies a minimum growth time of 140,000 years, assuming continuous growth conditions. The thickness of this reef was discovered when drilling holes for hydrogen bomb testing. The ideal growth rate is assumed to be that which has been observed. It may or may not be the case that the growth conditions currently observed are the ideal. A final conclusion can be elusive even for a scientist familiar with the measurements and assumptions.

GEOCHRONOLOGICAL QUESTIONS

There is an important distinction to be made between scientific theories and the empirical data with which they are associated. Facts and data are one thing; the interpretation of facts—that is, how to make sense of them—is another. It is often the case that a set of facts can be interpreted or construed in more than one manner. The few milestones in the heroic age of geology have shown how one interpretation came to be preferred over another. The raw facts of the rock layers, their geographic extent, ordering, thickness, material, and fossil content were organized in accordance with the uniformitarian view in order to devise an account of how these facts came to be. The theory was thereafter enhanced with the inclusion of catastrophic events to account for various initially nonconforming facts. Uniformitarian theory has continued to supply a reasonable and satisfying foundation for geochronology. But there are certain other facts that raise questions about it, including the following:

1. Fossil imprint and preservation can occur only with rapid burial. Absent rapid burial, the remains disintegrate, making recognizable fossil imprints impossible. This point is incompatible with geologic uniformitarianism and slow deposition of sediments. It would seem a rather important point to factor into any theory on the formation of fossil-bearing strata, especially considering the ubiquitousness of fossil-bearing strata around the globe. If the necessity of rapid burial is acknowledged at all in geology textbooks, it is not acknowledged in conjunction with uniformitarian accounts of the formation of fossil bearing strata.[11] The catastrophist position is a much better fit with fossil formation.

2. Carbon-14 is found in coal. The appearance of C-14 in coal is inconsistent with uniformitarian geochronology. The carboniferous period, during which, according to uniformitarian geology, many coal deposits were formed, was 280 to 345 million years ago. C-14 is also found in diamonds, which are formed well below the earth's surface under very specific conditions. But if coal deposits and diamond bearing formations are as old as uniformitarian geology says they are, with a half-life of 5,730 years, all C-14 should have long ago decayed to C-12.[12] The most straightforward explanation for the existence of C-14 in coal and diamonds is that they were formed within the range of C-14 dating applicability. To date, claims that the presence of C-14 is the result of contamination or underground radioactivity are conjectural.

These two points are among many arguments advanced against uniformitarian geology. Arguments that require a higher level of technical knowledge, unique findings that can be suppressed or explained away as anomalies and for which it is impractical or impossible to obtain independent verification, or arguments that amount to projections of a presently observed process into the past are omitted here.

THE OTHER THEORY REVISITED

To what should the rock layers be attributed? From the Bible's perspective, the globe-wide flood described in Genesis 6 through 8 is the answer. While the Bible says nothing about rock formations, this is the only event since the creation that could account for the globe-wide formation of massive sedimentary rock strata. How does that flood explain the stacked layers, and where did all the water come from?

Before proceeding to answer such questions, there is a need to acknowledge that with an event of this magnitude, in addition to being supernaturally caused, there are serious limits as to what can be empirically verified. Observations of catastrophic events occurring today, such as layered deposits formed during volcanic eruptions, are observations only of what has occurred today. Events that occurred in the unobserved past remain unobserved. The only way to explain rock formations of the unobserved past is to do what uniformitarian geologists do—attempt to reconstruct the past in accordance with a preconceived theory or else apply the same present-is-the-key-to-the-past approach but from a catastrophist perspective. Uniformitarian and catastrophist theories both rely on imagination because both have the same problems with empirical verification. On the basis of Genesis chronology and the record of the flood, the rock layers are explained as having been deposited by water, a supernaturally caused event that left observable effects. In uniformitarian geology, the rock layers are likewise explained as having been deposited in water, the observable effects of natural causes.

Without diminishing the striking similarity, there are two divergences between these theories. The first difference pertains to the rate with which the deposits accumulated and the time frame. On these points, the preservation of fossilized remains on a global scale should not be ignored. The evidence is that rapid burial was occurring, not just here and there, but on a vast scale.[13] The C-14 presence is evidence that at least some of the deposits were completed not nearly so long ago in the past.

The second difference is on how the root cause must be validated. Absent the Genesis chronology, it seems unlikely that formation of the rock strata in one globe-wide flood could ever be concluded from scientific evidence by itself. At least the history of the catastrophist view and its departure from the single-flood hypothesis shows that. Creation geochronology must therefore prove the veracity and authority of the Genesis record. And that it can do. (Chapter 11 presents an introduction to this subject.) Uniformitarian geochronology can only appeal to the nonrelevance of supernatural causes, which is not an assumption that can be scientifically proven.

CONCLUDING REMARKS

Geology is a science driven extensively by the economics of resource exploitation, and years of accumulated practical knowledge and associated theory have served it well. Few geologists have reason to question the assumptions of two centuries ago. But might it be possible that while the practical knowledge is sound, the explanatory theory is not? Are nonconforming facts being dismissed or disregarded? Why, for example, are meteorites not found buried in the sedimentary strata of the geologic column?[14]

Most of the creationist attention is focused on bringing down biological evolution. The arguments against evolutionary biology are more formidable than those against uniformitarian geochronology. This is probably why popular creationist opinion and even scientific opinion to some extent grant the age question to geology while rejecting biological evolution. However, unless the geochronology on which evolution rests is overthrown, the associated biology will remain standing. As a practical matter, the wide detour by way of the Bible that creationists must argue through cannot be sustained very well or even introduced at all in scientific circles. The divide that sets science and the Bible against each other is too well entrenched in the culture. There is a need, therefore, for the evidence and the assumptions used to justify uniformitarian geochronology to continue to be investigated.

Chapter 4:
Biology

*The "Origin" provided us with the working hypothesis we sought.
Moreover, it did the immense service of freeing us forever from the
dilemma—refuse to accept the creation hypothesis, and what have
you to propose that can be accepted by any cautious reasoner? In 1857,
I had no answer ready, and I do not think that anyone else had.*
                                            —Thomas H. Huxley
                                    (from a retrospective 1887 essay)

The 1859 publication of Charles Darwin's *Origin of Species* was one of
the great watersheds in the history of science. It redirected the science
of biology, utterly redefining the scientific understanding of the origin
of living things, and suddenly and substantially widened the heretofore
creeping divide between science and theology. It was because of this book
that the idea of biological evolution and the incremental developments
of living things spaced out over long periods of time became a scientific
theory. Though biological evolution existed as a philosophical idea prior
to Darwin, it had never gained acceptance as science. But when Darwin
correlated to it his detailed and systematic observations on adaptations of
certain species of animals, it quickly gained credibility as scientific. Within
a few decades of 1859, biological evolution as a scientific theory achieved
nearly complete acceptance by all scientists, not just biologists. From then
on, the fossil record and all scientific findings about living things were

organized in accordance with it. Yet for at least fifty years between 1809 and 1859, it was a theory in search of evidence.

A Theory in Search of Evidence

There are indications that the ancient Greeks had an idea about evolutionary progression in living things, though it was rather vaguely described as compared to the modern understanding of evolution.[1] The evolutionary idea seems to have disappeared after that, but it reemerged in the eighteenth century as a philosophy on the fringe of mainstream deism. Then, early in the nineteenth century, French naturalist Jean-Baptiste Pierre Antoine de Monet, Chevalier de Lamarck (1744–1829) made an attempt to gather up scattered ideas about evolution into a coherent scientific theory. In his 1809 book *Philosophie Zoologique*, Lamarck introduced an idea about species development known as inheritance of acquired characteristics.

Inheritance of acquired characteristics is the idea that modifications in individual organisms form as survival responses to their environments, that these are passed on to offspring, and that over time, this results in the formation of new species. Classic examples of this include giraffe necks and bat eyesight. The giraffe stretches its neck to reach leaves on trees and ends up with a slightly longer neck. This longer neck is passed on to its offspring. Each succeeding generation has slightly longer necks than the previous as the giraffes strain to reach higher and higher leaves. The eyesight of a bat, being extraneous to its navigation because it navigates by sound and because bats live in dark caves, is neglected and becomes weaker over the course of its life span. The weaker eyesight is passed onto the next generation until eventually, after many generations, the bat becomes completely sightless. *Philosophie Zoologique* lacked scientific rigor, as the idea that acquired characteristics could be inherited was speculative. No supporting evidence was offered, and the idea was not accepted as scientific. Lamarck is noted, however, for having devised the term *biology*.[2]

As uniformitarian and catastrophist geologies were competing in the early nineteenth century, French naturalist Georges Leopold Chretien Frederic Dagobert Cuvier (1769–1832) began theorizing about the developments of creatures preserved in the rock strata as fossils. He reasoned that any explanation of fossilized life-forms necessarily had to follow the same pattern as that of the rocks. His thinking was that organisms were suddenly caused to become extinct at various times in the past by unknown cataclysmic events, events that may also have changed the nature of subsequent geologic deposition. Following each of these events, other new and different types of organisms quickly appeared. This happened over and over again as the different types of fossils preserved in the successively stacked rock layers seemed to indicate. In order to correlate these observations with uniformitarian geology, the explanation of the fossils necessarily had to also presume long time periods where the forms of living things remained about the same. Cuvier maintained that species were fixed and did not evolve; he was opposed to Lamarck's hypothesis. Yet he also stressed how organisms had been adapted to their environments. Cuvier is regarded as the father of paleontology, the study of the fossil record.

In 1844, a volume titled *Vestiges of the Natural History of Creation* appeared, written by author and publisher Robert Chambers (1802–1871) but published anonymously. While Chambers was interested in getting evolutionary ideas out in public, he did not wish to have himself associated with ideas that were sure to be controversial. In *Vestiges*, living things were presumed to have resulted from the operation of laws of nature alone with no ongoing divine intervention. The exposition of a philosophy of evolution was the underlying motivation of this book, but as there were no empirical grounds, it was dismissed as unscientific. Even so, the idea of naturally occurring progress stirred the public imagination. *Vestiges* raised further awareness of the possibility of evolution in the public mind, and this effect was substantial. About this time, Charles Darwin was intending to publish an essay on evolution by natural selection, but seeing the unfavorable scientific reaction to *Vestiges*, he changed his mind. Clearly, scientific acceptance of evolution was going to need a serious appeal to evidence.

By the mid-nineteenth century, an ideology of technological and social progress began to emerge in Western culture. The achievements of the industrial age were a principle source of confidence in a human potential for technological and social improvement. This progress was characterized by the increasing specialization and ordered complexity seen in both machines and economic society. In 1844, an ideology of natural progress was emerging as well; the appearance of *Vestiges* is a clear indication that the idea of naturally occurring evolution was out and about and that its popularity was not justified on scientific merits. The idea of evolution was as much inherent to an ideology of technological and social progress as to an ideology of biological progress. The increasing specialization and ordered complexity of living things was viewed as the same sort of improvement. Darwin's evolutionary ideas were not hatched in a cultural vacuum.

Prior to 1859, the scientific consensus was that species of living things were fixed and did not evolve. Selective breeding was well understood and practiced at this time, but it was decidedly not considered grounds in support of evolution. The majority of naturalists accepted a divine creation, though one not specific to the book of Genesis. Divine creation was the only explanation available to them. And consistent with what Cuvier had theorized, multiple creations spaced out at specific intervals over long time periods seemed to be the best fit to what was known about the fossil record. Meanwhile, scientific observations on the characteristics of both living organisms and fossilized remains were accumulating and would eventually be used to support biological evolution. But a mechanism that could link the adaptations of organisms with the idea of upward progress in nature was missing.

Enter Charles Darwin

Charles Darwin (1809–1882) was taking his time as he was developing his theories and documenting the supporting evidence. He had first thought of the idea of natural selection in 1838 and had written the unpublished essay on that subject in 1844. He recognized that environmental factors and competitive pressures within a population could be involved in a selection process leading to the formation of new species. But when he saw the scientific hostility toward *Vestiges*, he realized that extensive references to empirical findings would be necessary. In 1858, he became aware of some competition. Naturalist Alfred Russel Wallace (1823–1913) sent him a manuscript wherein he had independently conceived of a process of species development that was similar to Darwin's. He had heard that Darwin was also working on the species problem and was interested in getting his opinion. The title of Wallace's manuscript was *On the Tendency of Varieties to Depart Indefinitely from the Original Type.*

Because the concept known as natural selection could now no longer be kept hidden, it was arranged for the Linnean Society in England to publish shortly thereafter (still in 1858) manuscripts from both men. Even though Wallace's ideas regarding species development were not nearly as well backed with supporting evidence or as carefully thought through as Darwin's, once Darwin realized the existence of competition, he got down to business. In late 1859, he published a first edition of *On the Origin of Species*.[3] Detailed and systematic observations of armadillos, tortoises, and finches obtained during the HMS *Beagle* voyage of 1831 to 1836 were his chief empirical evidence in support of the theory of natural selection.

Darwin had two main conclusions: (1) Natural selection of randomly emerging characteristics of organisms results from environmental pressures and determines what characteristics are passed on to subsequent generations, and (2) species are attributable to branching lineages from a common ancestry. The theory of natural selection says that variations in organisms arise randomly or accidentally in offspring that may or may not survive depending on which variations are best suited to the environment and best enable an organism to compete within a population. The organisms that survive pass on their variations to their offspring, and the variations are thus preserved through future generations, assuming they remain advantageous in the environment. Adaptation to environmental change is thus said to be a driver of species change over time. The mechanism of organism variability, however, as Darwin acknowledged, was unknown.

A second main conclusion was on the form of evolutionary progression. The formation of new species is the result of variations built upon variations over successive generations, a continual branching out and compounding of variations as they are found to be sustainable in the environment. This is presumed to be the explanation of the huge number and diversity of species that now exists. This point is often misunderstood. As an example of the correct understanding, birds and reptiles evolved from the same ancestral species, though one that does not now exist. The misconception is in thinking that one species existing today evolved from another species existing today, that birds evolved from reptiles; a fine point perhaps, but an important one for Darwin and evolutionary biologists in their quest to establish the idea that incremental changes over many generations could produce wide-ranging characteristics.

The most important aspect of *Origin of Species* and what made Darwin's work so influential is the extent to which he was able to support his conclusions with evidence. He called upon the abundant and detailed observations on the characteristics of various species of organisms that he had obtained during the *Beagle* voyage. One class of examples often cited were the various shapes and sizes of finch beaks that Darwin observed to be adapted to the type of food available in specific environments. It was this reference to abundant empirical data that gained evolution an appearance of scientific respectability that it had previously lacked. Nevertheless, the data simply indicated that selective breeding was being applied by natural environments; all the finches he had observed, for example, remained finches. That a species could transform or transition into a form of life that would have been considered a different species[4] was obviously an extrapolation of the data, and he acknowledged that too.

THE DARWINIAN REVOLUTION

The reception of Darwin's publication is noteworthy. Most initial scientific reactions were negative as was popular opinion. That *Origin of Species* raised serious ethical and theological implications was immediately understood, and it raised these implications to a degree to which no other scientific theory had ever come close. The hypothesis that all species of living things were the result of natural factors alone was in direct confrontation with the understanding of design inherent in miraculous creation. In 1859, Darwin had not mentioned anything about human evolution. (His other book, *The Descent of Man*, was not published until 1871.) But the logical connection from animals to human beings was inevitable. Despite its intensity, the initial controversy evaporated rather quickly. This was partly because of the new avenues of inquiry *Origin of Species* opened to science and partly because of its fit with the ideology of social and technological progress already mentioned. Another factor is that the theological problems were resolved to the satisfaction of many by invoking the idea of theistic evolution, that God somehow used evolution as the means of creation.

A further noteworthy development was in regard to the scientific acceptance of Darwin's mechanism of natural selection as the driver of evolutionary change. Within a few decades following *Origin of Species*, natural selection had been thoroughly rejected by scientists because it suggested nothing about how variations in species actually arose. Other mechanisms of change replaced natural selection within the evolutionary framework, notably Lamarck's inheritance of acquired characteristics, which had gained no further scientific credibility since 1809. Another mechanism was orthogenesis, which explained variations as being predisposed to certain directions by factors inherent to organisms. Biological evolution became widely accepted because of the scientific credibility implied by Darwin's data and analysis. Yet the immediate conclusion from his data and analysis was rejected, and what had previously been rejected as scientifically untenable (Lamarck's hypothesis) took its place. Rather a strange development! Evolutionary theory, once established, now seemed to thrive on ideas that were long ago rejected as unscientific. Not until the 1920s did natural selection begin to make a comeback.

In the decades following 1859, the chief advocate of Darwin's evolutionary theory was Thomas Henry Huxley (1825–95). Darwin's personality was such that he was disinclined to argue in support of his theory in public. Huxley had no such reticence. Even more than promoting Darwin's natural selection idea, Huxley effectively used the evolution hypothesis to force a phase-out of all appeals to supernatural and miraculous causes in biology and all science. Huxley and other naturalists had desired some hypothesis besides creation as evident in this excerpt from a retrospective 1887 essay of his:

> That which we were looking for, and could not find, was a hypothesis respecting the origin of known organic forms, which assumed the operation of no causes but such as could be proved to be actually at work. The 'Origin' provided us with the working hypothesis we sought. Moreover, it did the immense service of freeing us forever from the dilemma—refuse to accept the creation hypothesis, and what have you to propose that can be accepted by any cautious reasoner? In 1857, I had no answer ready, and I do not think that anyone else had.[5]

Darwin had supplied the missing hypothesis in *Origin of Species*. From the beginning, Huxley doubted that Darwin's natural selection was the cause of evolution, and sooner or later, most other nineteenth-century scientists also did. But he promoted the methodology employed by Darwin, his detailed and systematic observation, which gave the concept of evolution a scientific respectability it did not previously have. Due in large measure to his championing the methodology and his forceful expulsion of supernatural explanations in science, the long-standing negativity among scientists toward evolution was not only neutralized but transformed into acceptance and support of it as a scientific reality.

The significance of the change that Huxley helped to bring about in the understanding of what science is cannot be overstated. This change marked a final stage in an ongoing transition from science being descriptive of the natural world to explaining its origin as well. Uniformitarianism and the sufficiency of natural law were being worked for all their worth here. Scientists were now claiming jurisdiction that previously had been the domain of theology, the supernatural, and the God of creation; they were claiming jurisdiction over origin. As science displaced the supernatural explanation with a natural one, it was also displacing theology as the intellectual center of Western civilization. Science was becoming *the* path to knowledge, and because scientific inquiry into the supernatural is not possible, the intellectual respectability of inquiry into the nature of God was falling from favor. It was Huxley who invented the term "agnostic" to express the inability of scientific inquiry to determine whether God existed or even anything at all about the nature of God.

THE NEO-DARWINIAN SYNTHESIS

Through the 1920s, 1930s, and into the 1940s, biologists' comprehension of genetic principles led to a restoration of the supremacy of natural selection in the biological evolutionary framework. Because natural selection by itself could add no new characteristics and could only screen existing ones, it needed a mechanism capable of producing variations in species. The rediscovery of Gregor Mendel's experiments was key. Mendel (1822–1884) had experimented with how traits are inherited by offspring from their parents. His now-famous experiments with pea plants between 1856 and 1866 had lapsed into obscurity for many decades. But now in the twentieth century, their value was seen. With Mendel's genetic principles as a start, biologists began to obtain a working knowledge on how variations in the characteristics of organisms arose. This eventually led to the neo-Darwinian synthesis.

The neo-Darwinian synthesis united natural selection with genetics; it consisted of natural selection operating through environmental pressures on variations produced by principles of heredity. This synthesis came to maturity in the 1940s. The discovery of DNA and random, unpredictable mutations of genetic codes in 1953 resulted in an even more satisfactory overall theory in the 1950s. Knowledge of the role of genetics in determining how living organisms pass on characteristics and how physical characteristics may be changed by mutations greatly increased the confidence of scientists that evolution was indeed the explanation of all living things.

The theory of biological evolution has since been regarded as the unifying concept of biology. From geneticist Theodosius Dobzhansky (1900–1975), a chief formulator of the synthesis between genetics and natural selection, we have the following title of a 1972 paper: "Nothing in Biology Makes Sense Except in the Light of Evolution." The theory continues to gain explanatory sophistication as it is enhanced to account for new scientific discoveries about living organisms as they are made. The ever-increasing collection of fossilized remains also continues to be organized in accordance with evolutionary thinking. The structures, adaptations, and geographic distributions of organisms, living and fossilized, are all being explained by reference to natural selection operating on random mutation, the local isolation of species, and changing environments. All this correlation and organization of the data ever more strongly suggests to scientists that evolution really did take place.

REVISING THE SOCIAL ORDER

Confidence in scientific progress toward a comprehensive evolutionary theory was such that evolutionary thought began to be applied to explain the development of societies and cultures. The trend of ideas is not neutral in this regard; intellectual preference seems to lean toward removal of the transcendent. If biology followed this path, why should not social theory as well? In 1933, a philosophical and social doctrine known as secular humanism (also often termed atheistic humanism) was announced as the replacement for religion and what religion supplies to the culture.[6] Julian Huxley (1887–1975), a grandson of Thomas Huxley, and John Dewey (1859–1952) were advocates of this line of thinking along with several others of perceived intellectual stature. Evolutionary science served as secular humanism's foundation in the belief that mankind is the exclusive product of biological evolution. And if mankind's origin was evolutionary, so too were societies and cultures. Thus, all appeal to the transcendent and transcendent values and standards of behavior was targeted for removal.

An evolutionary concept of origin is a philosophical presupposition for which there are certain logical implications. It means that ideas about value, ethics, morality, right and wrong, and good and evil must have evolved just as living things did. All such ideas as well as those about God and the supernatural are presupposed to have been generated within the context of culture to keep a social order intact, to keep people from eliminating each other in the quest for survival of the fittest. This means that God is no more than a cultural device or mechanism—a further adaptation for survival on a community level. Because the behavioral restraints that religion supplies were seen to be a necessary element of civil society, secular humanism officially offered the rational mind as the replacement for religious faith. By reason alone, the new humanists surmised, matters of value, ethics, and morality could be sorted out. No longer would it be necessary to threaten people with divine justice to keep them in compliance.

The social and cultural reorientation proposed by the secular humanists seems as deliberate as any social and cultural movement that preceded it. But that deliberateness raises an interesting question: Was secular humanism to be just another stage of evolutionary development, or in keeping with the deliberate character, was there something more involved in its introduction than what may be accounted for by evolutionary theory? If another stage of development, should it be necessary to announce it, and what could be gained by doing so? If the latter, it would seem that evolutionary theory is not the total explanatory theory it is held up to be. Where intent comes from is missing from the theory.

The late 1950s saw a federally funded initiative to modernize science curriculums in public schools. By the early 1960s, the scientific approach to knowledge was becoming well positioned in public education. Further reorientation toward a secular knowledge base continued in the 1960s as federal courts sought to impose a new tack in the separation of church and state on public education.[7] Largely through public education, the conscious secular worldview of a few has become the unconscious philosophy of many to the point where secular humanism has now become the *de facto* philosophy and religion of our culture. Concurrent with the changes in education, shifts in public values regarding marriage, family, what constitutes an honorable lifestyle, and even life itself have become evident. The logical implications of a belief regarding one's origin tend to show up sooner or later. Though it is not so much that secular indoctrination has caused such shifts; rather, it has removed restraints on already existing tendencies toward such shifts.[8]

It may be worthwhile to ask to what degree evolution (the scientific theory) influenced the cultural ideology of technological and social progress versus to what degree the cultural ideology influenced the science. Historians have dealt with this issue, but to arrive at a definitive answer is difficult because of the speed with which new ideas and trends emerge and propagate. Positions are taken at both extremes as well as balanced approaches, but historians generally acknowledge that the interchange of ideas over the past two centuries has been two-way. Of particular note is one book by evolutionist historian and philosopher of science Michael Ruse (1940–) wherein he argues that the success of evolutionary theory can only be understood in the social context of the nineteenth and early twentieth centuries, in which it arose and prospered. Historian Peter Bowler notes of Ruse, "He too believes that the scientific credentials of the theory of evolution were never strong enough to convince anyone not predisposed to accept it on ideological grounds."[9]

REAL BIOLOGY CAUSES SOME DIFFICULTIES

The scientific theory of biological evolution runs up against certain problems that are seldom mentioned or admitted by scientists. Perhaps this is understandable, given that the trend and focus of scientific inquiry is toward increasing specialization on various and minute details of the theory. Awareness of the lack of empirical support that evolution can occur or did occur is lost. Following are what are generally accounted as problem areas for the theory. They have been known since Darwin's time:

1. Not a single empirically supported or even plausible mechanism for producing life from nonlife has been forthcoming either at the biochemical level or otherwise. The closest thing to a mechanism yet proposed is a casual appeal to the sufficiency of random interactions of molecules to produce life. Nothing approaching the rigor of a probability analysis is ever used to lend credibility to the supposed transformation of nonlife to life. A law known as the law of biogenesis consistently prevails—it takes life to get life.

2. While adaptations of species to their environment is an observable fact, neither transitions from species to different species nor increases in complexity have been observed in nature or produced by experiment. Mutations do not explain this because they are either inconsequential or degenerative. Only limited variations within species are ever observed in natural environments or produced in experiments. The assortment of claims that consist of nothing more than variations within species must be dismissed. A black-peppered moth is still as much a peppered moth as a lighter colored peppered moth; a large beaked finch is still as much a finch as a small beaked finch. In no case have compounded variations within a population, say, of finches, ever produced a creature that would have to be called "not a finch."

In explaining the evolution of living things—both their origination as well as subsequent development—scientists sometimes appeal to randomness in nature, and at other times, they appeal to laws of nature. But whether an appeal is made to randomness or to natural law, insurmountable difficulties arise. (See Appendix A.) Two further problem areas have been identified more recently and are associated with the intelligent design movement: irreducible complexity and the information content of DNA. These are described in chapter 7.

CONCLUDING REMARKS

I t may be noted that there is a continual emphasis within the science of evolution on progression from simple to complex over time. Seldom, if ever, is anything said about evolutionary regression, incremental decreasing complexity, or devolution. Yet there are plenty of lower order life-forms at least as successful at survival as more complex forms. Should not random mutations and natural selection be just as likely to produce devolution as evolution? The silence on decreasing complexity seems to confirm once again that evolution remains a philosophical idea, exerting a serious constraining influence upon scientific thought.

Biological evolutionary theory is a working explanation of life origin and development. It is a narrative of empirically inaccessible past events constrained by an overarching principle that nonnatural causes are strictly forbidden. Evolutionary biology was hatched in the nest of geological uniformitarianism. It was then reared within a philosophical and cultural environment that exalted the idea that technological and social progress, and by association biological progress, is inevitable. Later, as the neo-Darwinian synthesis inspired greater scientific confidence, evolutionary biology served as a prime justification for revising the social order. However, a total lack of empirical evidence to support the claim that natural causes are sufficient to explain the origin and development of living things would seem to render evolutionary biology unqualified for status as scientific knowledge. The problem is that evolution is the only possible *scientific* hypothesis. For science to give it up would be to relinquish its claim over biological origins to nonscientific or supernatural explanations. It does not appear likely that science will ever call a retreat in that regard.

It is now fair to say that science, having begun as a specialized method of inquiry on a limited scope of problems, has, in the confidence of its success, breached the boundaries of empirical verifiability it had set for itself, even judging as unscientific the very theology that attended its birth, the theology that says that nature is rationally ordered according to the purposes of its creator.

Chapter 5:
Astronomy and Physics

If the heavens above can be measured, and the foundations of the earth below can be explored, then I will cast off all the descendants of Israel for all that they have done, says the LORD.
—Jeremiah 31:37 RSV

Astronomy has presented its own challenges to the biblical record of creation, and so the history of this branch of science is also relevant. Besides the big bang theory and the age of the universe it implies, of particular interest is an answer to the following question: How is it possible to see so much and at such great distances in the universe if it is no more than several thousand years old? Unlike the challenges generated by geochronology and evolutionary biology, which are dependent on philosophical assumptions, this question is surely scientific, a matter involving measurements of speed and distance.

In contrast to the histories of geology and biology, the history of astronomy is more intriguing, with more turns, upsets, and overthrows. This history shows science at its best—entrenched and treasured theories cast aside because they could not be squared with observations or stand in the presence of superior theoretical insights. The trail of progress is strewn with discarded conceptions and intuitions about how the heavens are arranged and how they behave. We will consider only a few of these turns in continuing our historical survey. The history of astronomy is intertwined with that of physics, and we need to consider that history too, particularly the insights of Albert Einstein.

THE ORIGINAL UNIVERSE AND ITS FIRST REVISION

The ancient Greeks knew that the Earth was a sphere. They had observed its curved shadow on the Moon during a lunar eclipse and had recognized that the view of the stars was different at different latitudes. They were able to measure the angle of the shadow cast by a vertical pole at two different latitudes and obtain a fairly accurate estimate of the Earth's circumference and diameter. Eratosthenes made these measurements and calculations in the third century BC. The ancient Greeks successfully estimated the size of the Moon and calculated its distance from the Earth. This feat was achieved by Aristarchus, also in the third century BC.[1] Aristarchus also attempted to estimate the size of the Sun and its distance from Earth, though not successfully. He severely underestimated both measurements because the technology necessary to obtain even approximate estimates did not exist in the ancient world. But Aristarchus seriously considered the possibility that the Sun, not the Earth, was the center of the universe.

The idea that the Earth might not be the center of the universe did not get very far in the ancient world. The Earth just seems stationary; it is difficult to give that up. The ancient Greeks understood that if the Earth was moving in an orbit around the Sun, you should at least be able to perceive movement in the positions of nearer stars against other more remote stars from the different vantage points along the Earth's orbit. It turns out that no such motion can be detected with the unaided eye, nor could it have been detected with the technology available in the ancient world; the stars are simply too far away. Aristarchus had, in fact, suggested that the stars were too far away to see any shift in their positions, but there were no means for anyone to test that hypothesis. It was not until the mid-nineteenth century that improvements in telescopes reached a stage that enabled such motion to begin to be detected. The apparent attraction of all objects toward the center of the Earth was another compelling advantage of the geocentric universe. In addition to a lack of evidence of a moving Earth and the heliocentric configuration running counter to intuition, Plato and Aristotle, two of the most influential thinkers in the ancient world, had already taught that the Earth was the center of the universe.

In the second century AD, the Egyptian astronomer Claudius Ptolemy devised a surprisingly accurate geocentric mathematical model in order to plot and predict the motions of the five known planets: Mercury, Venus, Mars, Jupiter, and Saturn. This model consisted of circular orbits for each planet whose centers were slightly offset from Earth. Each of the five orbits had superimposed epicycles—that is, smaller circular orbits that proceeded along the main orbits—in order to explain the observed retrograde motions of the planets (the term *planet* means wanderer) and variations in their speed and brightness as viewed from Earth. Each of the five orbiting systems might be described as a big carousel, slightly off-centered and wobbly, on which sat a miniature carousel on which the planet itself sat. The five systems had different sizes and ran at different speeds. The Moon and Sun each had its own wobbly carousel too, though there was no need of mini-carousels for them. The overall visualization tends to get quite complicated, but the model achieved the objective of predicting planetary motion quite effectively. Islamic mathematicians and astronomers revised and improved Ptolemy's model from the eighth through the fourteenth centuries, but they never questioned that the Earth was properly at the center of all planetary motion.

Still, there was something about the complexity and arbitrary character of the Ptolemaic model that just did not seem right. It was also inconsistent with the natural philosophy of Aristotle that imagined the heavenly spheres to be of a simplicity and elegant perfection in their form and motion. In his 1543 *De Revolutionibus Orbium Coelestrium* (*On the Revolutions of the Heavenly Spheres*), Nicolaus Copernicus (1473–1543) proposed a heliocentric universe that was mathematically simpler than Ptolemy's model and eliminated the arbitrary complexity. Even though his mathematical model was slightly less accurate in predicting the positions of the planets than Ptolemy's revised and improved model, Copernicus's sun-centered hypothesis became known throughout sixteenth-century Europe. Its merits and demerits were seriously discussed and debated in academic circles. It also came to be appreciated and used by some astronomers. But because the geocentric view was intuitive and so well entrenched, it prevailed well into the next century. The fascinating aspect of Copernicus's idea is that it was conceived on mathematical grounds alone, not by attempting to form a model that might be a better fit to observations. This fundamentally new mathematical model of planetary motion was a deduction from Copernicus's premise that the universe was the work of a supremely good and orderly creator, an expert craftsman. He could not square the archaic model of Ptolemy with that premise. The heliocentric model satisfied the need for a more rational and intelligible universe, and it turned out to be right!

Tycho Brahe (1546–1601) was an outstanding early astronomer who amassed a large collection of observations on the trajectories of the five planets over many years in the late sixteenth century. Brahe's own hypothesis was that Earth was stationary and that the other planets orbited the Sun, which, in turn, orbited Earth. His collection of observations was well leveraged by Johannes Kepler (1571–1630), who mathematically solved a crucial puzzle of planetary motion in his 1609 work *Astronomia Nova* (*The New Astronomy*). Even though Kepler relied on Brahe's observations of planetary motion, he shared Copernicus's view on the arrangement of the universe. He had already made this point known in his earlier 1596 *Mysterium Cosmographicum* (*The Cosmographic Mystery*). Brahe's collection included particularly abundant and precise observations on the orbit of Mars, and Kepler found that these observations could be best explained if the orbit of Mars around the Sun was slightly elliptical. He further found that with elliptical planetary orbits, the accuracy of Copernicus's heliocentric model could be improved to the point that it became superior to the accuracy of Ptolemy's revised and improved geocentric model.

Kepler proposed certain laws of planetary motion in *Astronomia Nova*: (1) Planetary orbits, including Earth's, are ellipses with the Sun positioned at one of the two foci of each planet's ellipse, and (2) on the elliptical plane with the Sun as one point and the planet as the other, the area covered by a straight line between these two points on the elliptical plane over equal intervals of time is equal. This means that a planet speeds up as its orbit draws it nearer the Sun and slows down as its orbit draws it farther away. He later proposed a third law linking the time for a planet to complete an orbit to its average distance from the Sun. But even with the superior accuracy achieved by introducing elliptical orbits, the overall mathematical simplicity, and Galileo's discoveries by telescope at about the same time (see chapter 1), the case for a heliocentric universe was still not altogether convincing. It was not until 1687 when Isaac Newton's theory of gravitation in *Philosophiae Naturalis Principia Mathematica* (*Mathematical Principles of Natural Philosophy*), removed another obstacle to the belief that the planets, including Earth, orbited the Sun that the Sun-centered view was finally vindicated. Newton explained why objects fall to Earth: it is because of the gravitational property of mass, not because of anything to do with Earth being the center of the universe.

Kepler's achievements coupled mathematics with observations on the orbital paths of the planets. The Sun-centered model thus stepped beyond mere mathematical appeal. Both Kepler's and Galileo's contributions established the necessity of linking observations to any hypothesis or theory about physical reality. On the theoretical side, Newton's theory of gravity supplied backing for the validity of Kepler's laws of motion. Kepler's laws were, in fact, deducible from Newton's universal law of gravitation. We take for granted that science and mathematics are linked, that the natural world is rational and understandable by the rigorous logic of mathematical formulas. But in the seventeenth century, finding that there was such a linkage was an astonishing discovery.

But did Kepler, Galileo, and Newton really prove that the heliocentric view was correct? The answer is yes with the stipulation "based on what is known at the present time." This was and remains an inferred fact; it cannot be directly perceived. We should more properly say that the weight of evidence and logical argument had shifted in favor of the Sun. The laws of gravity and motion did not support the archaic motion required by the geocentric model. But how much confidence was there in the newly devised laws? Scientific observations and experiments in the seventeenth century were establishing the validity of gravitation. In addition, Galileo had realized that constant uniform motion cannot be perceived and that this fact explained why no one felt the Earth's axial rotation or its motion through space. Thus, reasons that were formerly advanced for maintaining an earth-centered view (i.e., no perceived motion, all objects attracted to Earth, and no essential correlation between mathematics and physical reality) were swept away. Today, we also know how much more massive the Sun is in comparison to any planet and that the inability to perceive changes in the positions of stars against other more remote stars at different locations in the Earth's orbit was because they were too far away for seventeenth-century instruments. But science continues to explore and discover; there is reasonable but never complete assurance that new findings will never change the inferences.

BEYOND THE SOLAR SYSTEM

In 1671, the distance between the Earth and the Sun was logically inferred from parallax measurements of the distance between Earth and Mars. As improvements in telescopes were made over the course of the next two centuries, it then became possible to use Earth's orbit around the Sun to triangulate the distance to several stars near the solar system.

In 1838, Friedrich Wilhelm Bessel, a German astronomer, obtained the first estimated distance to a star, though he was about 10 percent too short. That star is known as 61 Cygni, which today is measured at 11.4 light-years from Earth. The measurement was achieved using the extremes of Earth's orbit around the Sun to compare the positions of that star from the two different vantage points against the background of other more remote stars. Assuming a fixed position of the background stars, a star whose distance is being measured forms the far point of a triangle while the diameter of Earth's orbit defines one side of the triangle, the baseline. The angle of view against the background stars is compared six months apart and the length of the other two sides is determined using the geometry of triangles. The use of a known baseline in this manner, such as the extremes of Earth's orbit, is called parallax. Parallax is a triangulation method of measuring distance similar to the technique surveyors use in measuring distance. With parallax measurements of nearby stars, it was understood that stars were none other than "suns" whose brightness was diminished by their great distance from Earth. The successful use of parallax was the first step enabling astronomers to probe the depths of space.

Parallax is the most accurate and reliable method for measuring distance in space. It was the only reliable means of estimating distance to objects in space until the 1920s. The technique is limited because of the imprecision involved in measuring extremely small angles, angles less than one arc-second or less than 1/360th of a degree. The difficulty of these measurements may be appreciated in that, as of 1908, only about a hundred stars had had their distances measured by parallax.[2] By the late 1970s, parallax distances of the less than five hundred stars within a range of about seventy light-years[3] had been measured by ground-based telescopes. The best ground-based telescopes today have a practical parallax range of about three hundred light-years. However, with orbiting telescopes, the number of stars for which parallax measurements have been obtained has risen to about 120,000.[4] These 120,000 stars are within a radius of about 1,400 light-years, which represents a volume of space about one hundred times larger.

Attempting to figure out how far away any particular star is beyond the range of parallax is very confusing. How bright a star appears is generally the only means for doing so if it is too far away for the parallax technique to work. The apparent brightness of an object is inversely proportional to the square of its distance. But a dim star relatively close by could have the same apparent brightness as a brighter star that is farther away, and there is no way to tell the difference between them. An initial assumption that all stars have approximately the same brightness was soon shown to be invalid by the catalogue of parallax measurements being slowly built up in the late nineteenth and early twentieth centuries. Brightness versus parallax distance was not turning out to be a uniform function. A few other more approximate, more complicated, and less reliable techniques for measuring distance were devised in the late nineteenth and early twentieth centuries, but we will not cover these here.

There were other objects in space besides planets and stars. Fuzzy patches of light had been noticed in space since the time of ancient Greece. These fuzzy patches were called nebulae from the Latin root of the English term *nebulous* meaning unclear or indistinct. They were similar in appearance to comets but were distinguished from comets in that they did not move. We take for granted today that these are galaxies as the Milky Way is a galaxy, but before the 1920s, no one was sure what they were. Astronomer and master telescope builder William Herschell, who had discovered the planet Uranus in 1781, proposed that the Milky Way was all there was to the universe and that these fuzzy items were a part of the Milky Way and might possibly be new stars being formed. In contrast, Immanuel Kant, a German philosopher, imagined these nebulae to be star systems similar to the Milky Way itself and extremely far away. In that same year (1781), French astronomer Charles Messier published a catalogue of the known nebulae to prevent them from being confused with comets. It remained a mystery, however, as to what these objects were, whether they were indeed part of the Milky Way star system or far outside it. Improvements in telescopes began to unravel the mystery, but it was not until the mid-1920s that it was solved completely.

EINSTEIN AND RELATIVITY THEORY

Because astronomers depend on light in their study of objects in space, it may be a good idea that they know something about it. By late in the nineteenth century, the speed of light had been measured very close to the presently measured value of 299,792,458 meters per second (approximately 186,000 miles per second). But a huge scientific question toward the end of that century was this: What was this speed with respect to? Was this speed relative to any observer anywhere, or was it with respect to some absolute motionless frame of reference within the universe? Having already shown by experiment that light could travel in a vacuum—that is, a space devoid of atoms and molecules—physicists in the nineteenth century were convinced of the existence of *ether*, an unobserved substance with special properties that permeated the entire universe. Ether was necessary to provide a medium for the propagation of electromagnetic radiation, which includes light as well as radio, microwave, ultraviolet, X-ray, gamma ray, and infrared radiation. They thought that because air serves as the medium of propagation for sound and is the frame of reference for measuring the speed of sound, light also must have a medium and frame of reference. There was only one problem with ether: Its existence was sheer conjecture.

In 1905, a German physicist named Albert Einstein (1879–1955) stood the world of science on its head with his theory of special relativity. He adopted a straightforward principle much like Copernicus did that revolutionized the scientific understanding and mathematical description of the universe. His theory extended what was known about some laws of physics, the laws of motion, and applied that to other laws of physics, particularly the laws of electricity and magnetism. In doing this, he resolved a fundamental contradiction between laws of motion and laws of electricity and magnetism. This is because until 1905, the laws of motion were considered invariant to any observer, whether that observer was himself in motion or not, but the laws of electricity and magnetism were considered to be invariant only with respect to an absolute frame of reference. If the laws governing electricity and magnetism were indeed invariant only with respect to an absolute motionless frame of reference, these laws would be different on Earth as compared to elsewhere in the solar system or universe. An experiment conducted in 1887 failed to detect Earth's motion with respect to this presumed absolute frame of reference (the supposed ether), and this left scientists baffled. The experiment was done to compare the speed of light traveling in line with Earth's motion through space with the speed of light perpendicular to Earth's motion. This experiment, known as the Michelson-Morley experiment, was beyond reproach, skillfully designed to remove all possible influences that might hinder detection of the ether. It was not detected. Still, no one could accept that there was not yet some absolute reference for electromagnetic radiation. Einstein resolved this puzzle and the contradiction between laws of motion and laws of electromagnetism with the theory of special relativity.

Galileo had first recognized that one's perception of motion is relative. Take, as a simple example, a train car. If the car's motion is perfectly smooth, without looking out a window, you will not be able to tell if the train is moving or not. If you were to walk back and forth inside the train car while it is moving, you would not perceive any difference between that and walking back and forth in the car while the train is stopped. If you were to drop something on the floor, you would see it drop the same way with the train moving as you would see it drop if the train were stopped. If you were to pour hot coffee, you would pour it in precisely the same manner with the train moving smoothly along as you would if it were stopped. You would not need to make adjustments to account for the train's motion. This illustrates the relativity principle as applied to laws of motion. The term relative is used here because there is no absolute frame of reference to which the laws of motion are bound. The physical laws governing motion inside the train car remain exactly the same to you whether the train is moving or not.

Galileo further understood that even while the Earth revolves around the Sun at great speed and also rotates on its axis, this motion is not perceived. As long as any motion is steady, there is no means to perceive that motion. This was Galileo's principle of relativity. The only motion that counts is what is relative to an observer, and there is no absolute frame of reference against which motion must be measured. This principle had answered at least one objection to a sun-centered universe: why no one felt the Earth moving. Constant motion within a frame of reference is perceived the same regardless of whether or not the frame of reference itself is moving. Our frame of reference for motion is Earth. But we could certainly imagine other frames of reference whose motion would be different than Earth's—any of the planets, comets, other galaxies, the Moon, or even just the train. But in all these cases, assuming no change in gravity, we would expect the laws of motion to remain the same.

Einstein applied Galileo's principle of relativity to say that *all* physical laws governing the universe are perceived to be the same by any observer of them. Einstein said that the laws governing electricity and magnetism, which apply to light and the above mentioned forms of radiation, are also perceived in the same way regardless of where or under what conditions of motion they are observed. An observer's own motion makes no difference. As with laws governing motion, Einstein claimed there is no absolute frame of reference from which these other laws are referenced; they are relative to the observer's frame of reference, not to some absolute motionless frame of reference like ether. A first version of relativity theory, special relativity (special in the sense of specialized or limited) was presented in a 1905 paper, and by 1915 (published in 1916), Einstein had completed a more generally applicable formulation of relativity theory called general relativity that even superseded Newton's law of gravity.

The term *relativity* to identify this theory can be misleading if you infer from that term that everything is relative and nothing is absolute. It does not mean that at all. It means that physical laws are absolute—so absolute that there is no variability to them no matter what your point of view might be, whether on Earth or speeding through space at a substantial fraction of the speed of light with respect to Earth. What ends up being relative are your perceptions of time and distance, dimensions previously thought to be absolute and universal. Einstein did not name his theory relativity, and he did not like the term relativity; he would rather have had his theory known as the theory of invariance.[5]

Einstein did not conduct any physical experiments; his insights were exercises in thought. His first exercise in thought began with this question: If he himself were traveling forward approaching the speed of light, would he be able to see his face in a mirror he was holding out in front of him?[6] His answer was yes. He would still perceive light to be traveling between his face and the mirror at three hundred thousand kilometers per second. By this, he would *not* be able to tell whether his own speed was approaching the speed of light or if he was even moving at all. The experimental evidence of Michelson and Morley had shown that the measured speed of light is not affected by the frame of reference in which it is measured. So in order to remove the contradiction in the physical laws, the speed of light had to be constant with respect to any observer—that is, if a source of light were moving toward you or moving away from you, the measured speed of light had to still be the same. And if the speed of light was the same with respect to any observer, regardless of how fast or slow the observer was moving, then the speed of light must be a universal physical constant. His first deduction from the constancy of the speed of light was that because speed is simply distance divided by time, distance and time were not the absolute measures everyone thought they were. He made this theoretical leap when all other scientists remained stuck on an absolute frame of reference for electromagnetic radiation. They were all intent on explaining away the failure of the Michelson-Morley experiment to validate an absolute frame of reference (perhaps the ether somehow moved along with Earth or perhaps Earth was compressed as it moved through the ether to just the right degree so that the speed of light ended up appearing the same in both parallel and perpendicular directions).

A NON-EUCLIDEAN UNIVERSE

Relativity theory begins with a straightforward idea but leads to profound and disturbing implications on the nature of time and distance. The implications of relativity theory are that everyone's intuitive understanding of time and distance as absolute and universal is wrong. Intuitively, time seems absolute as well as universal. We would say the same of distance and spacial dimensions. This is not true. The passage of time and the measurement of distance become relative to the observer. Two observers moving at different speeds will view time and distance differently. From the viewpoint of a stationary observer, time runs slow within the reference frame of a second observer who has accelerated to a significant fraction of the speed of light. From the viewpoint of the accelerated observer, time runs fast within the reference frame of the observer who has not accelerated. But within his own frame of reference, the accelerated observer perceives no change in the passage of time. This effect of relativity is often characterized as decelerated aging of space travelers; they show up on Earth after having dashed about in space over what seems a short time and find that everyone they knew has long since been dead.

The effect on distance is that it shrinks from the viewpoint of a stationary observer viewing an accelerated observer. The distance a stationary observer measures within the reference frame of an accelerated observer is shortened along the direction of motion. An example typically used to illustrate this is the perceived length of a plane or a spaceship as it is speeding by. The length perceived by a stationary observer becomes shortened, and it approaches zero length as the speed of light is approached. This shrinkage is required in order for the speed of light to remain a constant. If there was no shrinkage, a beam of light sent from the back of the space ship to the front would be measured faster than three hundred thousand meters per second. That would, of course, be a contradiction. The shrinkage effect is consistent with the time effect because the passage of time approaches a state of stopped or frozen time as the speed of light is approached. Time becomes frozen in light as it were, and so any distance traversed by light, no matter how great, becomes no distance at all because no time has elapsed. An even stranger effect is that events that one observer says occur at the same time occur at different times as viewed by another observer traveling at a fraction of the speed of light compared to the first observer. These effects are minute until the speeds being discussed become significant fractions of the speed of light, but the effects become extreme as the speed of light is approached. A difficulty in validating these effects on time and distance lies in a third effect: As you try to accelerate any object to approach the speed of light, its inertia increases. Inertia increases so much, in fact, that it becomes impossible to accelerate any object to the speed of light because its inertia approaches infinity.[7]

Gravity produces effects that are the same as those produced by acceleration on the perception and measurement of time and distance. As gravitational fields increase in intensity, time again slows down, and distance decreases. Time theoretically approaches a stopped state, and the perception of distance shrinks toward zero as you encounter gravitational fields strong enough to bend light back on itself such as in a black hole. Einstein realized that there is no recognizable difference between gravity and acceleration; the force of a gravitational field is equivalent to the force perceived while one is in a state of acceleration or deceleration. And because the perception of gravity disappears in a state of free fall, it is not a universal physical law applicable to all frames of reference. This led Einstein to conclude that gravity is not so much a force as Newton had described it but rather a non-Euclidean geometry of space and time in the vicinity of massive objects. Massive objects warp space (and time) so that space is no longer Euclidean; the shortest distance between two points is no longer a straight line and two parallel lines can intersect. Earth is not massive enough to cause a noticeable influence on the curvature of space. But non-Euclidean space becomes noticeable in the vicinity of objects with extremely high mass like the Sun. And the Sun is an object of low mass compared with many other objects out in space.

In 1919, British physicist Sir Arthur Eddington (1882–1944) led an expedition to test and compare Einstein's and Newton's gravitational theories during a solar eclipse. The test involved measuring the amount of deflection of starlight by the Sun's gravity. This can be accomplished during a total solar eclipse because stars that cannot be seen during the day become visible. If Einstein was correct, stars that were behind the Sun would be seen because the Sun's gravity would bend the light rays enough to curve slightly around it. They would appear as being beside the Sun in the same direction of sight. Eddington's expedition observed and measured the bending of light around the Sun and found that this was in accordance with what Einstein's theory was predicting and that this bending was not as well predicted by Isaac Newton's theory. Other observations and experiments have since further verified that Einstein's theory is more correct than Newton's. Newton's 1687 theory of gravity is not so much replaced as it is improved upon in that Einstein's theory explains what is observed under more conditions than Newton's. Newton's theory is accurate in low gravitational fields, such as those on Earth, for most applications. In strong gravitational fields, it is necessary to use Einstein's theory.

In a 1971 experiment, atomic clocks were flown around the world and compared with clocks that were not flown. A difference in time was noted, and though extremely small, the discrepancy was consistent with general relativity theory. The difference was attributed to the relative motion between the flown and stationary clocks and the effect of reduced gravitation from the higher altitude on the flown clocks.[8] Other observations have validated the theory, including further observations on the bending of starlight in high gravitational fields and observations of distance traveled by short-lived subatomic particles. As an example of the theory's application, the GPS (global positioning system) must take relativity into account in sending and receiving signals from the array of GPS satellites. If adjustments for gravitational effects on time are not made, the system will not supply accurate locations and will be in error by a matter of miles.[9]

Here again, a new theory on physical reality, counterintuitive and in defiance of common sense, has replaced that which was considered an absolute fact. Relativity theory has radically altered science's perceptions of the universe. The theory's counterintuitive character and difficulty in comprehension is why a certain amount of controversy over it remains. But the lack of intuitive comprehension is only because the effects are beyond our experience; they only become significant under high gravitational fields or at speeds approaching the speed of light. In the same fashion as Copernicus, Einstein was convinced that the universe must be more ordered and rational than it appeared to be from the state of physics at the end of the nineteenth century, and he pressed on with this view on philosophical grounds. The strange implications are all rigorous logical derivations from the premise of uniform and consistent laws of motion *and* electromagnetics. So is relativity a proven fact? Again, the answer is yes with the stipulation "based on what is known at present." The mathematical rigor of the formulas of relativity theory and the rational appeal of unified laws of physics are forceful arguments that it is true. The accumulating observations and experiments, while spare, also validate its correctness. But again, scientific knowledge is never final; it is not inconceivable that future discoveries might expand the knowledge base such that relativity theory might need to be replaced by something that explains more than what relativity does now. Einstein himself was candid about this possibility.

Sir Arthur Eddington, through his advocacy of relativity, came to be closely associated with it in the public mind. Physicist Ludwig Silberstein once said to Eddington, "You must be one of the three persons in the world who understands general relativity." Eddington remained silent. Silberstein said he didn't need to be so modest, to which Eddington replied, "On the contrary, I am trying to think of who the third person might be."[10] This overview is given so at least you can be aware of the theory and some of its consequences. It would appear to be essential in understanding what goes on in the universe. Further explanation is required beyond what has been presented here in order to understand the theory and its implications, but there are good tutorials available for anyone inclined to pursue the subject further.

MEASURING DISTANCE IN REMOTE SPACE

From 1894 through 1912, Henrietta Leavitt (1868–1921), a volunteer at Harvard College Observatory, was cataloguing and studying photographic plates of stars taken by telescopes with particular attention to stars whose brightness varied over time. A certain class of stars whose brightness varies are known as Cepheids after an earlier discovery that the star known as Delta Cephei has a variable brightness. She eventually discovered about 2,400 of these variable stars. Leavitt hoped to find some linkage between average brightness and the time period of variation in brightness that could be used as a means for estimating their distance. Generally, it was impossible to be sure how far away any of these variable stars were because there was no independent means to measure their distances. The nearest variable star, Polaris, the North Star, was still too far away to have its distance measured by parallax.

But there was an idea that could be used to get around this problem. Her studies of variable stars in the Small Magellanic Cloud, a cluster of stars seen in the Southern Hemisphere, might at least indicate a relative relationship between average brightness and the period of variation. The distance to the Small Magellanic Cloud was unknown at that time, but it was considered far enough away that all stars within it might reasonably be considered to be about the same distance from Earth. She discovered twenty-five of these variable stars in the Small Magellanic Cloud. She found that the time intervals or periods for one cycle of the variation in brightness to occur ran from being as short as a day to as long as a month and that this time interval was proportional to the average brightness observed. The variable stars in the Small Magellanic Cloud could be used to estimate distances of other nebulae in proportion to the Small Magellanic Cloud if variable stars could be found in them. Later in the 1920s, it became possible to estimate the distance to one of these variable stars (though not one in the Small Magellanic Cloud and not by parallax). The result was an approximate means of measuring the distance to other variable stars.

Early in the twentieth century, certain questions were being asked: Was the Milky Way all there was to the universe? How big was it? Was the solar system at the center of Milky Way or closer to its edge? Were nebulae such as Andromeda part of the Milky Way or separate? Answers to these questions were the subject of serious debate among astronomers. In 1920, a debate was staged between two American astronomers over the question of whether the Milky Way was all there was to the universe, the nebulae being just parts of it. Or were the nebulae other "Milky Ways" at huge distances from our own galaxy? Harlow Shapley argued that the Milky Way was the entire universe and that nebulae were small satellites nearby. Heber Curtis argued the other position that nebulae were comparable to the Milky Way and at great distances away. But Shapley also argued that our Sun is far from the center of the Milky Way, while Curtis argued that it is right at the center of it. The debate was inconclusive. In 1920, there was a near total lack of empirical evidence that could be called in support of any of these positions.

In 1912, American astronomer Vesto Slipher (1875–1969), in his work at the Lowell Observatory in Arizona, began to apply the science of spectroscopy to the light from the mysterious nebulae. Spectroscopy was developed in the nineteenth century and is the means by which astronomers can analyze the color spectrums of light to determine things like temperature and chemical composition of the source of light. Knowing the composition, it is then possible to determine the radial speed of a source of light from shifts in the expected frequencies in the light spectrum. A shift in frequency due to speed is known as a Doppler shift after Christian Doppler, who theorized in 1842 that waves would appear compressed or expanded as the wave source moved toward or away from an observer. This effect is similar to a familiar effect on sound waves: The pitch of sound is higher when the sound source is moving toward you and lower if moving away. With light, a redshift (lower frequency) is observed when a light source is moving away, and a blueshift (higher frequency) is observed if the light source is moving closer. This shifting does not affect the speed of light, only its wavelengths. Slipher observed red and blue shifting but mostly red in light from the nebulae. A redshift meant the nebulae were moving away from Earth, and a blueshift meant they were moving toward Earth. By 1914, he had acquired the Doppler shifts of fifteen spiral nebulae, all but two of which were redshifted. He continued this work and eventually catalogued forty-one of these nebulae and their speeds. With four additions by other astronomers, a total of forty-five of these nebulae had been documented by 1925, and all but two of these had redshifted light spectrums. This was about as many nebulae as the twenty-four-inch refracting telescope at Lowell and other telescopes could reach in 1925. It was still not known what these nebulae were or how far away any of them were. But the speeds being measured were as high as a thousand kilometers per second, one third of one percent of the speed of light.

Soon after he published his theory of general relativity in 1916, Einstein realized that the theory implied a universe that was not static. It had to either be expanding or contracting, anything except remaining the same. It had to be expanding in order to counteract the tendency of gravity to pull everything together, or else it had to be collapsing because of gravity. This consequence, he thought, was preposterous; certainly, the Milky Way was neither expanding nor contracting, and the Milky Way was most, if not all of the known universe. (Einstein was as yet unaware of Slipher's initial observations as there was generally little contact between physicists and astronomers at that point. Advances in astronomy were proceeding in the United States while the forefront of theoretical physics was headquartered in Europe.) In a 1917 paper, Einstein added a term to the general relativity theory to allow it to show that the universe could be static, all the stars and everything else simply remaining still in their places. Einstein inserted this "cosmological constant" (fudge factor) into his general relativity theory, absolutely convinced that the universe was neither expanding nor contracting.

Einstein's theory was being studied by other scientists and mathematicians. In 1922, Alexander Friedmann, a Russian mathematician, concluded from Einstein's theory that the universe could conceivably be doing all kinds of things, expanding, contracting, or existing in a state of asymptotically decreasing expansion, with or without the aid of the cosmological constant. Einstein, who was now a science celebrity, resented the suggestion that his theory could be interpreted in so many different ways and expressed harsh disagreement. He had expected that his theory would describe the universe only as it existed—static. It is true that his theory has many possible outcomes, but it was inconceivable to anyone at that time that the universe could be other than stationary because it just looked stationary. Indeed, astronomers had assumed the universe to be static in the absence of any evidence to the contrary.

Georges-Henri Lemaitre, a Belgian mathematician and physicist, independently reached the same conclusions as Friedmann between 1925 and 1927. In 1927, the redshifted nebulae findings of Slipher and others were published with assistance from Eddington. Though the identity of and distances to the nebulae whose redshifts had been measured were still unknown, Lemaitre went further and worked out an expanding universe solution to the general relativity equations that would approximately produce the level of redshifting that had been observed. This solution suggested that all matter had been compressed into a small area at a beginning state in time. This was a preview of sorts to the current big bang theory of the universe. Einstein agreed that these conclusions were mathematically sound, but he remained vigorously opposed to the idea of an expanding universe. The expanding universe hypothesis was not taken seriously by the scientific community either because empirical data needed to definitively support this or any other of the possible solutions to the general relativity equations did not exist.

About the point that Slipher had exhausted the possibilities of obtaining further spectroscopy observations at Lowell, distances to some of the mysterious nebulae began to be obtained. American astronomer Edwin Hubble (1889–1953), beginning in 1923 while at the Mount Wilson Observatory in California, was the first to make use of the Cepheids that Henrietta Leavitt had first proposed could be a key to measuring distances beyond those achievable by parallax. He identified Cepheid variable stars in the M31 (Andromeda) nebula and another nebula called NGC6822. He concluded that these nebulae were in fact separate, very remote galaxies. Based on the period of their variations in brightness, the Cepheids he had observed were about seven thousand times as bright as the Sun and indicated from the average apparent brightness that the nebulae were about nine and seven hundred thousand light-years away respectively. These findings were published in a 1925 paper. (The initial estimates have since been improved, and Andromeda is now measured by astronomers to be 2.25 million light-years away.) Andromeda and a few other nearby galaxies were close enough to be able to discern Cepheids within them, but other galaxies were too far away. The upper limit of Cepheid visibility with the

100-inch Mt. Wilson reflecting telescope was about ten million light-years. Hubble assumed that the brightest stars, stars much brighter still than Cepheids, in all remote nebulae were of the same absolute brightness, and he applied that assumption to estimate the distances to other galaxies too far away for Cepheids to be discerned. This assumption was justified by another assumption that there must be some physical upper limit to the size and brightness of stars without their immediate self-destruction.

The justification for this other assumption came out of another developing field of science, the science of astrophysics. Eddington and other physicists were attempting to solve the mystery of how stars produce light. Beginning in the 1920s, they combined known laws of gravitation, the physical properties of gases, atomic theory, and quantum physics (which was devised later in the 1920s and which explains interactions of subatomic particles) to show how stars could form and begin to generate thermonuclear reactions. This work eventually led to a theoretical understanding of how nuclear fusion occurs in the center of the Sun and other stars to convert hydrogen into helium. If stars could form out of gravitational compression of clouds of hydrogen gas, the temperature at the center of the cloud could become hot enough to produce nuclear fusion. Calculations on the rate of nuclear fusion indicated the rate at which hydrogen was converted into helium. This rate could then be used in estimating how long a star could exist and, from spectroscopic estimates of its composition, how far along any given star would be in its conversion of hydrogen to helium. From theoretical calculations, it was also understood that there could be physical limits to how large and bright a star could get without destroying itself in a huge explosion.

Distance measurement in remote space is difficult and imprecise. Beyond the range of parallax, distance measurements rely on apparent brightness of particular objects in comparison to their absolute brightness. The validity of assumptions on absolute brightness, sources of error in gravitational effects, dimming effects from space dust, and other factors need to be considered, which cannot be discussed here. To date, no Cepheid variable has had a direct parallax measurement of its distance so as to precisely calibrate their use to measure distances to other galaxies. This is because no Cepheid is close enough to Earth to have its distance measured by parallax. However, about twenty Cepheids are close enough to Earth to have their distance measured by a technique known as statistical parallax,[11] a more approximate means that will not be described here because of its technical nature. Correlations between spectra and brightness of stars have been used as distance indicators, but Cepheids are the most reliable distance indicators beyond the range of parallax. They happen to be extremely bright stars, and the distance to which they can still be detected is about fifteen million light-years with ground telescopes and about sixty million light-years with orbiting telescopes.[12]

Other means of estimating distances to galaxies too far away for individual stars to be seen have been devised more recently. Some of these involve supernovae. Supernovae are giant stars that have presumably exhausted the possibilities for further nuclear fusion and explode suddenly. These explosions are far brighter than the brightest normal star. Other means involve gravitational lensing of light from extremely remote objects and quasars. Collectively, the handful of distance-measuring techniques that have been devised have given astronomers some confidence in their distance measurements because the methods are independent of each other and have been found to achieve a degree of consistency, though not always to everyone's satisfaction. Astronomers routinely assign uncertainty bandwidths to all nonparallax distance measurements to reflect factors such as imprecision in measurement techniques, interference by intergalactic matter, and the range of uncertainty in the assumptions on which a particular distance measurement is based.

THE UNIVERSE IS REVISED AGAIN

While Hubble was estimating the distances to galaxies, Milton Humason, Hubble's assistant, also working with the 100-inch Mt. Wilson reflecting telescope, confirmed Slipher's redshift measurements and then began to obtain light spectrums of still more remote galaxies and their redshifts. By 1929, Hubble and Humason had gathered enough data to report an approximate trend linking the distance and redshift. The further away a galaxy was, the faster it appeared by the redshift measurement to be moving away from Earth. A galaxy that was twice as far away would on average exhibit twice as much redshifting. It could then be inferred to be moving twice as fast away from Earth. More confirming data of a redshift-distance proportionality was published in a 1931 paper. The scientific community endorsed these results. They were found to correlate with Einstein's general relativity theory through the work that Lemaitre had reported in 1927. The redshift-distance relation has since become known as Hubble's Law and the proportionality factor as the Hubble Parameter. It is noteworthy that Hubble reported his observed data only, allowing others to interpret from it an expanding universe and a point of time of its origin. Hubble did not ascribe the redshifts to the Doppler effect from galaxy recession either. He reported only what redshifting was observed.

In 1931, Einstein announced that he had abandoned the cosmological constant he had inserted into his theory in 1917 and turned to support an expanding universe. Later on, he said that the cosmological constant was the biggest mistake in his career. He further acknowledged that the most straightforward explanation for the redshifting of galactic light is that the galaxies were receding. Nevertheless, through the 1930s, astronomers and physicists generally clung to a static universe. While they agreed that Hubble's data showed the universe to be expanding, they argued that Hubble's data indicated the universe was younger than many of the stars in it.

With the idea of an expanding universe in hand, a team of physicists led by George Gamow, a Russian immigrant to the United States, devised what has come to be known as the big bang theory of the universe in the 1940s. This theory now proposes a universe expanding out of what is termed a singularity, a point in which all matter and energy occupied an infinitely small space at the beginning of time. In 1948, Gamow's team predicted that a luminous echo, an electromagnetic remnant of the "bang," should be observable throughout the universe. Astronomers made no effort to discover such an electromagnetic remnant, and this prediction was forgotten. Even at this point, physicists and astronomers were still working independently and did not customarily exchange ideas. Another team of physicists led by British physicist-astronomer Fred Hoyle (also in the 1940s) proposed a competing "steady state" model of the universe, one which also undergoes expansion, but with the difference that matter is somehow generated to fill in the expanding space such that a constant density of matter per volume of space is maintained.

Astronomers Walter Baade and Alan Sandage now began to revisit, revise, and expand on Hubble's data and to recalibrate the distance scale. Revisions to Hubble's measurements and the addition of many new measurements eventually placed the beginning of the universe further back in time as the distances of galaxies were found to be larger than first estimated by Hubble. Better estimates were made possible by better telescopes: the 200-inch Mt. Palomar telescope, completed in late 1947, now allowed clearer views into the depths of space than any telescope in existence to that point. By the mid-1950s, observations of thousands of galaxies, the discovery and study of various types of galaxies, and accumulating distance measurements within the Milky Way gave assurance to astronomers that the Milky Way was only one spiral galaxy out of many.

The development of radio astronomy, the study of radio waves in space, led to an accidental discovery that confirmed a prediction of the big bang theory. In 1965, American physicist-astronomers Arno Penzias and Robert Wilson, while attempting to calibrate a new type of radio telescope, found that no matter where they aimed it, a low-level "noise" was picked up in the microwave bandwidth of the electromagnetic spectrum. It was soon discovered that there was a certain frequency of microwave radiation coming from all directions of space. On consultation with other astronomers and physicists, this was concluded to be the predicted luminous echo from the universe's initial expansion. Light generated in the early universe had, it was theorized, stretched into the longer wavelengths of the microwave range of electromagnetic frequency as the universe expanded. The first of several satellites commissioned to chart this microwave radiation over the sphere of space was launched in 1992. The discovery of this microwave radiation is one of the main reasons Gamow's theory eclipsed Hoyle's steady-state theory and why the big bang theory remains so popular among scientists today.

Because of the difficulties of distance measurement, astronomers began to leave their distance-measuring techniques behind and plunged ahead in mapping the observable universe based on the redshift-distance relation alone. Redshift measurements of remote galaxies are more easily obtained than distance measurements. The temptation is to use the Hubble redshift-distance relation and assume distances to remote galaxies can be determined by their redshift alone. This introduces further sources of error because recession velocities are not solely a function of the universe's expansion but are also affected by the internal rotation of galaxies as well as the rotation of individual galaxies around gravitational centers of groups of galaxies. Besides these factors, the redshift-distance relation itself has been subject to continual adjustment as further distance measurements are added to the existing catalogue. The work of distance measurement continues in order to more precisely determine the value of the Hubble Parameter, which in turn gives the rate of expansion of the universe and thus, its age. Yet because of uncertainties in distance measurements in remote space, there continues to be a degree of dispute among astronomers even today over the value of the Hubble Parameter.

But is the universe really expanding? Again, the answer is yes with the stipulation "based on what is known at present." The expansion is inferred knowledge not directly perceived. The convergence of theory and observation toward a conclusion opposed to the static universe that astronomers and physicists had always maintained is a forceful argument that it is true. At the present time, there is nothing else being proposed that could explain the consistent pattern of observations in all directions linking the redshifting of light to distance that also satisfies the requirements of relativity theory. But again, scientific knowledge is never final; it is not inconceivable that future discoveries might enlarge the knowledge base such that the inferred expansion might need to be replaced by something that explains more than what expansion does now.

The big bang theory infers an initial "bang" from (1) an expanding universe, which is, in turn, inferred from measurements of redshift and distance, and (2) the principle of uninterrupted uniformity, which involves reversing (contracting) the observed expansion into the past back to a point in time when no further contraction is theoretically possible. The theory is thereby forced to posit a beginning of sorts to the universe, though scientists rather awkwardly find themselves unequipped to explain a beginning. The problem they are obliged to deal with is that of first causation. A question about the first cause for the expansion arises. But the singularity out of which the universe was supposedly formed is a point by definition where none of the laws of physics apply.

Since the big bang theory gained popularity in midcentury, scientific inquiry into how it explains the formation of galaxies, stars, planets, and the universe's distribution of galaxies has continued. More recent discoveries by astronomers are further challenging what in previous decades was assumed to be a sufficient explanation of the universe's formation. Theorists have repeatedly been driven to modify their theories in attempts to better explain new findings and to answer questions that have been raised more recently on how and why the universe displays certain characteristics and not others.

An example of such findings and questions involves a peculiarity in the rotation of spiral galaxies. Their rotation cannot be explained by the gravitational effects of the visible matter in them. The rotation of stars around a galaxy's center is observed to be at the same rate (number of revolutions per unit time) at the galaxy's edges as for stars near the center. This is in contrast to a solar system, where a planet's rate of rotation around its sun is proportional to its distance (Kepler's third law of planetary motion). The same type of rotation is observed in groups of galaxies where the individual galaxies revolve around a common central point.

A hypothesis about the existence of unseen matter has been advanced to explain this peculiarity in rotation. This matter is known as dark matter. There is only one problem with dark matter (a familiar one): Its existence is sheer conjecture. No one has ever seen dark matter, and no one has any idea of what it is. Experiments designed to detect it have failed to detect it. Yet it is believed to comprise as much as 90 to 96 percent of all matter. Such is the state of scientific knowledge that nothing is known about most of what is thought to comprise the universe. The dark matter hypothesis is reminiscent of the ether hypothesis. It would seem there are again serious disjunctions between observation and theory that await resolution.

Yet even with all the unanswered questions, the history of astronomy and physics has shown science at its best—the overthrow of entrenched paradigms and the confirmation of new and counterintuitive theories on the basis of empirical data and mathematical reasoning. A heliocentric universe supplanted a geocentric universe. A theory that could explain the propagation of light through empty space replaced a theory that could not. An expanding universe with a beginning displaced a stationary one with no apparent beginning. Some overthrows were led by the formulation of hypotheses, and others by unexpected observations; however, both theory and observation were necessary. The historical record indicates that observation is not all that is required to come up with a successful theory. The observed trajectories of the planets by themselves were fundamentally inexplicable without a sun-centered model and the laws of planetary motion and gravity. The redshift-distance observations by themselves were likewise inexplicable without the mathematical brilliance supplied by relativity theory. But then theory by itself is likewise insufficient. Kepler's laws of planetary motion and Einstein's relativity theory by themselves would seem to have no bearing on reality without the empirical evidence that confirmed them.

ESTIMATING THE AGE OF THE UNIVERSE

A static universe of limited size presented no particular challenge to a straightforward interpretation of Genesis 1. But what about an expanding universe of unimaginable size? The big bang theory seems consistent with creation in that a beginning is conceived. There may not be anything particularly conflicting about shifting spectral lines or the inferred expansion of space either. Expansion does not seem out of line with Old Testament phrasing that describes the heavens as being stretched out.[13] While no one ever conceived of anything resembling an expanding universe from the text of Genesis 1, there does not seem to be an obvious contradiction. There is an apparent conflict though between the big bang's uniformitarian approach to the universe's past and Genesis 1 on the question of how old the universe is.

Two lines of reasoning are used by astronomers to bound an estimate on the age of the universe from its apparent size: (1) Its age must be at least as old as the time necessary for light to arrive on Earth from the most distant objects seen in space, and (2) its age must be about equal to the time taken by galaxies remote from the Milky Way and moving away from it to have traveled that far away at their present rate of speed. Certain assumptions are required for these approaches to be valid:

1. The speed of light is a constant at all times and in all places.
2. Measurement of distance in space, while difficult and itself involving assumptions and inferences, yields approximately correct values of distance.
3. The observed redshift of the light spectra of remote galaxies is due to these light sources moving away or more properly, the expansion of space.
4. The universe is approximately as old as a regression in time of the presently observed rate of expansion for all matter to occupy a single point.[14]

None of these assumptions are provable, but they have each survived critical analysis. The speed of light cannot be changed without affecting other physical laws and upsetting the finely tuned balance of physical laws necessary for things such as life to exist. There is no evidence to suggest it has changed over time either. The second assumption has occupied much of the above historical review and has been and continues to be a primary occupation of astronomers. They have known that the reliability of their age estimates depends on the accuracy of their measurement of distances. A few independent methods of estimating distance beyond the range of parallax have been devised, and agreement between these different methods has been achieved to a degree. It is unreasonable to think that distance estimates are so inaccurate as to allow for the possibility that what is seen in space is all within a several-thousand-light-year radius.[15] The third assumption is the most straightforward explanation of the observed redshift and the only one that is consistent with both the known laws of physics and what is known about stars and their composition.[16] The fourth is a deduction from the uniformity principle, the principle that presently observed laws and operations of nature can be projected into the unobserved past.

A third line of reasoning on age involves the life cycle of individual stars. The age of the universe must be at least as old as the oldest stars. Star age is calculated from estimates on the rate of nuclear fusion and the supply of hydrogen and helium in a star's center. This approach also requires certain assumptions:

5. Physical laws governing nuclear fusion are constant at all times and in all places.
6. Astrophysicists' theoretical understanding of star life cycles is adequately representative of what goes on in the interior of stars.
7. Stars are approximately as old as a regression in time of the present rate of hydrogen-to-helium conversion to a total hydrogen composition.[17]

The fifth assumption has grounds similar to the constant speed of light. The sixth cannot be proven either but is checked in part by physicists' knowledge of nuclear bombs, laboratory experiments involving subatomic particles, and nuclear fusion, as well as observations of stars at different stages of their thermonuclear cycles. Stars in the Milky Way galaxy are observed in various phases from their supposed beginning as gravitationally compressed clouds of hot gas to their final phase as white dwarfs or neutron stars. The seventh is a deduction from the uniformity principle.

A most persistent and often asked question is how it is possible to see so much and at such great distances in the universe if it is no more than several thousand years old. Even the immediate galaxy itself presents this challenge. The Sun has been estimated to be about thirty thousand light-years from the center of the galaxy, and the galaxy is estimated to be about a hundred thousand light-years in diameter. The distance in light-years is also the look-back time, the amount of time taken by light to traverse the distance, and hence the elapsed time since the light being observed emanated from an observed object. The question does not so much involve the assumption of uniformity but simply the measurements of the speed of light and estimated distances of objects in space.

It has been proposed that objects more than several thousand light-years away are seen because their light-in-transit was created at the same time as the objects themselves and made to appear as if it were generated from the objects. The first problem is that this is an unverifiable proposition, and propositions that can never be verified cannot be regarded as knowledge. But even if that problem is excused, there still remains the fact that astronomers observe events happening at large distances in space—exploding stars known as novas or supernovas. If the speed of light has been finite and constant and the distances of millions and billions of light-years are even somewhat correct, these types of events should have happened in the distant past. One could say that the light-in-transit also was created underway from there to here. But that would seem to be crossing the line beyond what qualifies as an honest interpretation of what is observed because these events would not then have taken place. Events within a certain horizon measured in light-years since the creation could have taken place (i.e., a horizon of six to ten thousand light-years), but similar events just outside that horizon would not have. With regard to events in space, arbitrarily invoking deceptive appearances makes light-in-transit creation a difficult position to defend. Why should God create an illusory universe when He could fashion a real one anyway? "Is anything too hard for the LORD?"[18]

Changes in the laws or constants of nature have also been proposed in answer to the light time-distance question. The speed of light was faster in the past, say. If there is evidence of past change in a constant of nature, then fine. But if not (and it does not appear that there is), is a change in a constant being proposed solely to get around an otherwise intractable problem? It would seem out of character to suppose that the Creator would have had to modify or supersede constants of nature in creating, particularly because He would be the one who came up with the idea of having constants of nature in the first place. Arbitrary escapes from the problem of light time-distance do not get us closer to knowledge of the universe's age.

There is, however, a factor that should be relevant to how the age of the universe is measured and one that seems overlooked. Should that factor be respected, it might well nullify the light time-distance and uniformity puzzles, if not entirely putting the subject of age beyond the reach of scientific methods.

A Possibility of Alternate Theories

Alternate approaches to calculating elapsed time and age should be possible because of relativity effects on how time and distance are perceived. It is a wonder that nothing is said about gravitational effects on time and distance in past stages of the universe's presumably more compact existence. If the sense of absolute and universal time and distance was once shown to be incorrect, why is it still being applied? Or why is the Newtonian concept of time and space imposed on an Einsteinian universe? These questions seem to be answered by the theory described next; someone has thought to consider relativity effects.

An alternate theory on the universe's formation devised by physicist Russell Humphreys and published in 1994 considers extreme gravitational time dilation as the explanation of how light from remote galaxies arrived on Earth in a newly created universe.[19] This theory is similar to the big bang theory in that it also has an expanding universe, but it departs significantly in that it figures in relativistic effects on time and distance as the universe's mass expands out from what is known in relativistic terms as an event horizon. For a mass large enough and compact enough to where its gravity bends light back on itself, its event horizon is the boundary surrounding it and encompassing the space in which that bending back occurs. It is also the boundary inside of which time stops and outside of which time begins. In black-hole terms, it is the boundary of no return surrounding a black hole. But instead of a black hole wherein matter and energy are drawn in by gravity, Humphreys posits a white hole where matter and energy are emitted out. If the Earth was the last or near the last object to cross the event horizon on its way out, its clock would have begun only after the universe was essentially complete as seen today.[20] This theory also accounts for the cosmic microwave background radiation discovered by Penzias and Wilson that is often considered the definitive evidence of a big bang.

A chief scientific objection raised against such a theory would be its dependence on miraculous intervention. How is matter and energy supposed to overcome the gravity and get itself out across the event horizon? But that sort of objection is universal. The big bang theory also has miraculous dependencies. What is supposed to cause it in the first place—and from a singularity, a point wherein none of the laws of physics apply, and which is indistinguishable from nothing?

CONCLUDING REMARKS

I t cannot be said that all the laws of nature, the properties of the universe, or their ramifications under all conditions are known. Future discoveries and theoretical insights may push universe origin and formation theories in new directions that presently are inconceivable. A definitive answer on the formation of the universe and how the perceived look-back time might correlate with Genesis 1 chronology remains an open question. For now, one theory, that of Russell Humphreys, offers a correlation that seems as consistent with what is known by science today as the big bang theory. The main difference is that Humphrey's theory takes into account relativity effects on time, whereas the uniformitarian approach of the big bang does not.

The scale of the universe consistently humbles advances in the reach of observation. Imagination is overtaxed in conceiving a space-time continuum. Minimally validated theories are strained to describe phenomena in regions forever inaccessible to the experimental method. Astronomers say that looking out into space is looking back in time, so there is no way to know what is going on in remote space "right now." It would seem there are insurmountable spacial and temporal barriers to observation and discovery.

Outer space and the universe generate a lot of mystery and wonder, and a search for meaning. Yet after the first sentence of Genesis, attention is turned to the surface of the Earth with only brief recollections on the formation of the heavens in the remainder of the scriptures. They seem to be implying that the mystery, wonder, and meaning of existence is what happens on Earth, not outer space. I would venture that the mind that demands explanations for what it sees offers more insight into the meaning of existence than discoveries in the universe ever will.

Chapter 6:
Science 2

*What we learn from experience depends on the kind
of philosophy we bring to experience.*[1]

—C. S. Lewis

With a background in the history of the scientific revolution and that of geology, biology, and astronomy, it is time to attempt a summary appraisal of scientific claims about natural history and origins. We should be aware now not only of what scientists say about the origin of the earth, universe, and living things (the claim that they have undergone a long evolutionary history) but also the underlying assumptions—the sufficiency of natural causes and naturally occurring specialization and ordered complexity. But preliminary to an appraisal, some remarks on science in general as the means of acquiring knowledge of the natural world should be helpful.

SCIENCE IN GENERAL

What is science? An informal definition of science as currently practiced is the search for knowledge of what the natural world is, how it works, and how it got here—or how we can make sense out of what the natural world presents to us. More specifically, science is a procedure for generating hypotheses and testing them against reality. It seeks to identify, qualify, and quantify natural cause-and-effect relationships—that is, laws of nature. Hypotheses about the operations of cause and effect are confirmed or disproved by reference to empirical findings, and theories are generated when a set of related and confirmed hypotheses are widely accepted after they have undergone rigorous peer review. In science, theory development is the goal. Theories are the general truths, the sought-after explanations of what is observed in nature. A collection of facts by itself is useless until an organizing theory is found that explains them and continues to accommodate subsequently acquired facts. Astronomer and science writer Donald Goldsmith has remarked, "Facts become useful and significant only when we begin to recognize the organizing principles behind them. A good theory provides an organizing principle that one can carry through life and apply to situation after situation."[2]

What science aims toward, therefore, are theoretical concepts that organize, unify, and explain the raw data of facts, specifically scientific observations and experimental results. Science desires to find their underlying rationality and coherence. Theories and hypotheses also lead to questions of this sort: If the theory or hypothesis is correct, certain effects should follow, but do they? The urge to test a theory or hypothesis motivates scientific inquiry. So besides organizing, unifying, and explaining facts, a second reason why theories and hypotheses are important in science is that they generate a framework for continuing research and inquiry. Science could get nowhere without hypotheses and theories to test. Two further attributes that qualify a theory as a good scientific theory are (1) that it be successful in making predictions about the phenomenon under study—that is, given certain causes and conditions—what effects follow and (2) that it enjoy wide acceptance among scientists.

It may well be that a scientific theory has all four of these attributes (explains scientific facts, is testable, is successful in making predictions, and is widely accepted), but that cannot guarantee against future discoveries or insights necessitating its revision or replacement. The world of nature is complex. New discoveries in science confirm this point repeatedly. Hypotheses and theories depend on assumptions that are seen to be representative of the phenomenon under study within the bounds of current knowledge and observation. But as knowledge and observation expand, it is often the case that assumptions of the past can no longer stand as universal and must be replaced with others, perhaps less simplified and more interrelated, that better connect with newly discovered cause-and-effect phenomena. The claim that any scientific theory is *the* true and final account of the way the natural world really is problematic. Consider what Nobel Prize-winning physicist Arno Penzias says to a lay audience regarding theories of science:

> I originally came to science thinking that theories are proven or disproven. In actuality, they are accepted or abandoned. Scientists can't prove theories in absolute terms—"proof" comes down to practical experience. When scientists describe science to the public, though, we tend to present our stuff as truth. But we tend to forget underlying assumptions that are basically unprovable—such as, that the simplest theory is the right one. There's more of a human element in science than the public usually knows. Science is not so unreliable that any crazy idea is to be admitted—but it's a thin line, and a lot shakier than supporters of the scientific method would have you believe.[3]

Contrary to how science is typically portrayed in public, as knowledge of nature based solely on evidence, it is inevitable that human elements (assumptions) are present. And where there are assumptions, the possibility exists for alternate assumptions. Preference for simplicity, for example, is a factor that influences how the evidence is interpreted. The candor of Penzias's remarks is appreciated; we do not usually get as honest an assessment as this in a culture that views scientific methods as the only source of knowledge.

Theories should thus be regarded as tentative in that future discoveries may disprove them or cause them to be modified. As we have seen, there are instances in the history of science of widely accepted theories being discarded and replaced with ones that better explain the ever-expanding collection of observed facts. Examples of this would include elliptical planetary orbits replacing circular, Einstein's theory of gravity replacing Newton's, and the expanding universe theory replacing the stationary universe. Though scientists tend to regard observations and experimental results that confirm a theory (or hypothesis) as proof of its validity, such evidence cannot certify a theory as true with a capital "T." Particular observations and experimental outcomes cannot prove a theory making universal claims; it is not inconceivable that other theories built on alternate assumptions could make the same successful predictions.[4] In this, we see why it can never be said that a reigning theory will never end up on the scrap heap. Yet in the politics of science, entrenched theories can be used and often are used to marginalize and cripple nonconforming inquiry.

There are limits to what scientific inquiry can uncover. Ultimate "why" questions categorically cannot be answered, and answers to "how" questions are often beyond practical means of verification. Even so, science, as we have seen, restricts its scope of inquiry to natural causes. The problem with appeals to supernatural causes in science and why they are rejected is this: If supernatural causes were to be accepted, further inquiry into natural causes would cease. Supernatural causes are not susceptible to scientific procedures and should they be put forward as hypotheses, scientific investigation would have nowhere further to go.

Natural History and Origin

The quantity of scientific observations and findings now available and brought to bear on the subject area of natural history and origins is more than enough to overwhelm a nonscientist. As a practical matter, the nonscientist is dependent on the scientific community to make interpretations of the evidence. But then the manner in which a heap of evidence is organized and presented has much to do with what seems persuasive and convincing.

While it may be necessary to defer to scientists for the interpretation of evidence, the nonscientist is not incapable of forming considered appraisals. The power of scientific methods is not universal; it does not extend beyond the reach of empirical verifiability, even though science treads beyond that reach in some instances. Science is at its best only when it has access to observable and repeatable phenomena, conditions, and events. But in that what happened in the precivilized past was not observed by science, the first and best means of obtaining scientific knowledge about natural history and origins is removed.

It could be that natural history and origins are repeatable. Experiments may be able to demonstrate natural processes alleged to have been involved in natural history. But there is little or nothing to draw upon. Experiments exhibiting transitions from one species to another species or transitions from molecules to living cells have not been forthcoming. Neither has evolutionary development beyond variations within species been found operating in the world today. Theoretical processes thought to have been involved in the formation of the universe, the solar system, and planets remain beyond practical means of experimental verification. Some geological processes observed today may be a repetition of processes involved in forming the earth's rock strata, but what they produce, if they produce anything, is rather diminutive compared to the massive formations to be explained. Repeatability is generally either absent or unobtainable and therefore unhelpful in validating the natural histories devised by biology, geology, and astronomy.

Where observability and repeatability fail, it is still possible for scientists to propose theories. Scientific findings may be fit together into a preconceived picture much like one does with pieces of a jigsaw puzzle. With a puzzle, the picture on the front of the box is that picture. With science's constructs of natural history, the picture is constrained by this rule: Natural causes only need apply. Natural causes only effectively pictures natural histories according to the principle of uninterrupted uniformity. This is the principle that what is observed happening in the present is assumed to be what has generally happened in the unobserved past; it mirrors natural causes only.

As scientists have stipulated exclusive appeal to natural cause-and-effect processes in scientific theories, it is not surprising that biology, geology, and astronomy have produced natural history theories ruled by the uniformity principle. Geology and astronomy used it in devising a chronology of the earth and the universe. Biology added to that a second principle involving incremental development with increasing complexity and specialization over time. The past cultural appeal of the idea of progress associated with the industrial revolution and its interplay with evolutionary biology (noted in chapter 4) appears to have covered for the absence of evidence that incremental development with increasing complexity actually occurs over time. Biology then joined with paleontology, the study of the fossil record, to similarly devise a chronology of living things.

It may well be that the natural history and origin theories of biology, geology, and astronomy satisfy the four points noted above that qualify theories as scientific (explain scientific evidence, testable, successful in making predictions, and widely accepted). How close this is to thinking of them as true! Scientific origin theories are presented as knowledge of reality derived from evidence, but review of the histories of biology, geology, and astronomy shows otherwise and identifies the presence of assumptions that are not scientifically justifiable. Inevitably, the natural history and origin theories imply sufficiency to explain everything that exists in nature, and so the assumption of uniformity (alternately the stipulation of natural causes only) appears as if it were a conclusion. Again, we may have to defer to science for interpretations of the data. But then we should reserve the right to uncover and evaluate assumptions that rule the interpretation and presentation of that data.

It is true that uninterrupted uniformity cannot be discarded. It is expected that every observation and experimental test of nature will affirm that nature is rationally ordered and consistent. Time after time, we find that to be so. Projecting the uniformity of nature into the future becomes automatic. The record of the observed past is conclusive on the validity of this position, it is felt. Is there a limit as to how far uniformity may be projected into the unobserved past? And if so, where does that limit fall? Intellectual integrity demands that the limits of historical observation and empirical verification be acknowledged. The last three chapters have already noted reasons why the conventional uniformitarian stories of the unobserved past are suspect.[5]

In Perspective

The collection of evolutionary theories cannot be abandoned without forcing science to relinquish the territory of natural history and origin. Therefore, they persist. Yet the approach of explaining everything in terms of natural cause and effect is unsatisfactory on the whole. That approach can never supply ultimate answers to why, and answers as to how are often beyond any means of verification. Penzias continues:

> Moreover, no cosmological theory yet explains how or why the universe came into being. The universe seems to have an inherently unknowable—I won't say biblical—part to it. So we certainly can't say that everything's been done and there's no room left for wonder. We enter the twenty-first century with the same picture of the universe that George Gamow would have understood half a century ago, and we are still left with the mystery of existence. That, to me, is remarkable.[6]

The mystery of existence remains despite tremendous scientific progress. Should science even be expected to uncover the mystery of existence? And why should there be a sense of mystery in the first place? What is it about the mind that there should be a mystery? It appears that Aristotle's long overlooked final cause returns to haunt; the mind cannot entirely rid itself of the *why* of existence. Though if ever the reason for our existing becomes a quest, we should not expect science to supply an answer. Long ago, and as a precondition, science removed the *why*, Aristotle's purpose cause, from its scope of inquiry. That also is remarkable.

Chapter 7:
Controversy

Woe is me, my mother, that you bore me, a man of
strife and contention to the whole land! I have not lent,
nor have I borrowed, yet all of them curse me.
—Jeremiah 15:10 RSV

Anyone who is aware of the controversy surrounding the origin of the universe, earth, and living things should also know how the discord has been stationed squarely within public school biological science education. This is the sphere wherein the clash between evolution and creation has most engaged the public conscience about what is right and true. Parents, teachers, school administrators, judges, and government leaders all recognize the importance of science education to national prosperity and preeminence. But what should be taught in science classes regarding origins has generated often bitter controversy over nearly a century. The forces of contention in this war of ideas have played out in court cases involving the teaching of evolution and creation as science in public schools. How legal maneuvering appealing to the Establishment Clause of the First Amendment has excluded creation from public school science education becomes apparent from these court cases.

FROM 1859 TO 1925

To summarize developments leading up to the first clash over the content of biology classes, beginning in the 1860s, the controversy that *Origin of Species* stirred up was over whether Darwin's naturalistic account of species development should displace a purpose-driven theistic explanation. Darwin had dealt only with species development in 1859. He had not said anything about how living things first originated, and not until 1871 did he make any claims regarding the origin of human beings. Yet there were logical implications arising from Darwin's theory, and many followed that logic to an origination that removed God and purpose from the world. This, of course, was unacceptable to others, and hence, the controversy.

Darwin's natural selection hypothesis was soon set aside because there was no proposal as to the source of variability in living things upon which natural selection operates. Other hypotheses on variability were substituted into the new evolutionary science, specifically inheritance of acquired characteristics and orthogenesis. These substitute ideas were more compatible with purposeful design than natural selection and permitted a conjunction between theistic purpose and evolution. Various perspectives on theistic evolution became a subject of discussion and debate in scientific and theological circles. These perspectives tended to smooth out discord on the scientific-religious interface, and at least on the surface, they seemed to pacify the underlying incongruities between science and theology on the subject of origins. There arose a general unity of opinion through the last decades of the nineteenth century: Theistic evolution seemed the best overall fit between science and theology.

About this time, a new approach to understanding the Bible began to find a home in various Protestant denominations. This new approach, known as historical or higher criticism, was a theological program that began with a premise that the scriptures were of human origin rather than divinely inspired, and not surprisingly, ended up concluding that yes, they were of human origin—a fine exercise in circular reasoning. Given the premise of human origin, any occurrence of the miraculous was assumed to be categorically false. Deference to science was granted wherever science was treading, and it was found that respectability for religious belief could be purchased by purging it of supernatural content. The "solar system" mistake of the Roman Catholic Church, which, to this day, continues to be erroneously portrayed as *the* triumph of science over religious superstition, was particularly in mind. And so it became necessary to avoid being seen repeating clashes with science. This submission of theology to science was known by the term *modernism*.[1]

The intensity of the controversy that had boiled over in the 1860s had thus subsided. Scientists had embraced evolutionary theory because it offered them new avenues of inquiry; the public, whose attentions are generally focused on immediate threats to life and comfort, had forgotten about it. Evolution as a "survival of the fittest" philosophy of social progress and economic theory was entertained among the "enlightened," and biological and social evolutionary ideas were welcomed by a liberal clergy then taking charge of many Protestant denominations.

In the early twentieth century, certain developments in the United States led to a return of civic discord. The emergence of public secondary education—that is, high school education—in the two decades straddling the turn of the century was one. The teaching of science was a major part of the secondary education curriculum, and as such, so were botany and zoology. Botany and zoology were eventually combined into one program of study: biology. And since a large portion of the study of living things had by this time come to consist of evolutionary theory, evolution entered the schools.

Another early twentieth-century development involved a growing sub-trend among Protestant denominations stressing fundamental doctrines of Christian faith and a return to traditional interpretations of the scriptures. This was a reaction against the trend toward wholesale reinterpretation by historical criticism. This initiative to stress scriptural inerrancy and fundamental doctrines was also a reaction against the social Darwinism of the late nineteenth and early twentieth centuries. Social Darwinism was an evolutionary "survival of the fittest" worldview used to justify competitive business practices, economic exploitation, and ideologies such as German racial superiority and eugenics, the regulating of reproduction among "less desirable" members of the human population. This fundamental rejection of evolutionism (social, economic, and natural) insisted on a return to biblical literalism and Genesis creation in its miraculousness and purposeful design.

Still another development, particularly in the 1920s, was that advances in genetics had begun to allow for the gradual return of Darwin's natural selection to the evolutionary framework. As a result, evolutionary thought began to veer toward the naturalistic and away from the theistic. Natural selection through environmental factors was viewed as a thoroughly autonomous natural process, leaving no room for supernatural intent or purpose. This trend left the previous compromise of theistic evolution with an ever-widening gap to bridge.

The Scopes Trial of 1925

In the early 1920s, fundamentalist-inspired objections began to be raised against the teaching of evolution in the public schools. Citizens, especially in southern states, publicly opposed educational ideas with which they did not agree and that were subversive to their children's religious upbringing. Between 1923 and 1928, legislatures in four states, Arkansas, Mississippi, Oklahoma, and Tennessee, enacted laws restricting or prohibiting the teaching of evolution in public schools. Louisiana and Texas imposed administrative rulings against evolutionary content in textbooks. The Florida legislature passed a nonbinding resolution against teaching evolution. Local school boards throughout the South also imposed their own restrictions on teaching evolution. In the state of Tennessee, the antievolution law was known as the Butler Act of 1925. The statute was enacted on the principle that the electorate, acting through their legislature, has the right to determine what teachers, as state employees, should teach in the public schools. Even so, the measure was generally understood to be a symbolic one that would not be enforced. It was signed into law by the state's governor, Austin Peay, with this understanding.

Indignant over these state laws, the recently formed New York City-based American Civil Liberties Union (ACLU)[2] determined to test their constitutionality. The ACLU viewed them as impediments to academic freedom and freedom of speech: "How could state legislatures prohibit scientific truths from being taught?" The Butler Act was of particular interest to the ACLU because it specifically prohibited the teaching of any theory that denied the creation of man as taught in the Bible.[3] This, they surmised, would surely stand as an infraction against the separation of church and state. The ACLU advertised in Tennessee newspapers in an attempt to solicit volunteer science teachers who would agree to be charged with teaching evolution. In Dayton, Tennessee, some local businessmen took up the ACLU's offer and made arrangements with John T. Scopes, a twenty-four-year-old science teacher and football coach at Rhea Central High School. Scopes agreed to be charged with teaching evolution. It should be noted that the ACLU was interested in concocting a constitutional test case. They did not want legal proceedings to turn into a media event that could run afoul of that objective. However, the local businessmen had purposes of their own in mind; they were interested in promoting and publicizing the town of Dayton and thought that the attention a trial over such a controversial matter might bring would accomplish that objective.

The ACLU's strategy of creating a narrow constitutional test case was further compromised when William Jennings Bryan offered his services to the legal prosecution team free of charge. Bryan was a former congressman, three times a Democratic presidential candidate, and secretary of state under President Woodrow Wilson. He was a popular and able politician, a defender of working-class rights, and a conservative Christian. It was Bryan who had largely initiated the antievolution campaign responsible for generating the public support needed for passing the antievolution laws in the first place. Once Bryan jumped in, well-known defense attorney and agnostic Clarence Darrow, a vigorous opponent of Mr. Bryan's religious ideals, could not resist the challenge and immediately offered his services to the legal defense team, also free of charge. The ACLU attorneys at first accepted, though some later had second thoughts regarding Darrow's participation, not wanting to turn their test case into a science-versus-religion debate. They merely wanted to challenge the restriction on academic freedom. However, in short time, the ACLU found itself with little control over the unfolding of events.[4]

The press throughout the nation took full advantage of the looming spectacle anticipated in the squaring off of these two accomplished, though antagonistic orators. Everything possible was done to arouse national interest in the evolution-creation controversy. Speeches and debates on the merits and demerits of evolution and Genesis creation were staged before and after the trial. Articles and books were written. Controversial questions of science versus religion, academic freedom versus parental rights, and majority rule versus individual liberty supplied further catalysts for generating public interest. William Jennings Bryan took on evolution for the cause of Christian conservatism; Clarence Darrow, defending evolution in the name of academic freedom, took on biblical literalism. Opportunistic reporting was typified by H. L. Mencken of the *Baltimore Sun*, already well-established as a lead critic and ridiculer of religious fundamentalism. The role of the media in making the "Scopes Monkey Trial" the "trial of the century" was primary.

The eight-day trial spanned mid-July of 1925 in a hot and crowded Rhea County courthouse amid street vendors, monkeys, and sensationalized publicity around the nation. The prosecution undertook to restrict its case to a simple violation of a law on the books while simultaneously defending the right of states to regulate public education. They wished to steer clear of the larger controversies over evolution-creation and academic freedom. The defense's strategy was to demonstrate that evolution was not at all incompatible with religious faith, that evolution was a scientific fact, and that the Butler Act was an unconstitutional preference toward a particularly narrow religious viewpoint.

The state of Tennessee's case against Scopes was completed quickly. Much of the next several days were occupied by arguments over the validity and interpretation of the Butler Act and whether expert testimony should be admitted. In a first victory for the prosecution, the validity of the Butler Act was sustained by Judge John Raulston on the grounds that regulation of state-funded schools by the state legislature was within its legal jurisdiction. Academic freedom was not the issue because anyone could teach whatever they wished outside school walls. On the question of interpretation, the court upheld the prosecution's argument that the intent of the act was clear and needed no expert clarification. The defense had gathered various expert witnesses, both scientists and theologians, to testify on the validity of evolution as a scientific theory and how it was compatible with the religious beliefs of many. This expert testimony was excluded from the courtroom by Judge Raulston on the basis that it was not relevant to whether or not the law had been violated. It was submitted, however, as written testimony for the record. Scopes himself was never called to testify. There was a desire on both sides to avoid revealing any uncertainty about whether he had even taught evolution; he was not the biology teacher, and he had only once substituted for the regular biology teacher.[5]

Seeing their arguments on legal challenges to the Butler Act refused, the defense undertook to publicly ridicule a literal interpretation of Genesis creation and religiously motivated suppression of academic freedom. In a clever strategic move by Darrow, Bryan was called to the witness stand to testify on his beliefs regarding the Bible. With the permission of Judge Raulston and Bryan's consent, questioning by Darrow was allowed to proceed on the afternoon of the seventh day, a Monday. Bryan, though a capable debater and vigorous defender of the Bible, was placing himself at a disadvantage as a witness under examination. This was clearly a surprise move by Darrow, and Bryan's eagerness to defend the faith resulted in an impulsive decision contrary to the prosecution's strategy of steering clear of larger issues. It was also about to lead to one of the most extraordinary courtroom exchanges in legal history.[6]

What sorts of questions were asked by Darrow? The questioning forced Bryan to choose between natural and supernatural explanations, or for a precisely defined boundary between them: "Did Jonah live inside a whale for three days? Do you believe Joshua made the sun stand still? What would happen to the earth if it stopped moving?" Or he was asked to answer what can only be answered by conjecture: "Where did Cain get his wife?" Or he was compelled to choose between literal and figurative interpretations: "Have you any idea of the length of these periods (in reference to the days of Genesis 1)? Do you think the sun was made on the fourth day? Was there evening and morning without the sun?" The record indicates that Bryan was careful about defining terms being used, distinguishing between literal and figurative language, and was effective in questioning the credibility of scientific evidence contradicting the Bible.[7]

Yet while on the stand during the relentless questioning, Bryan took one position that turned out to be regrettable: when questioned about the meaning of the six days of creation, he stated that it was possible to view them as periods of time, not necessarily days. Protestant theology had long since conceded the age question to science, and that influence was apparent in his answers. This was regrettable because in a very public way, it raised the possibility of figurative interpretations of nonfigurative passages and the idea that the Bible can legitimately be interpreted in multiple ways. The cross-examination was abruptly halted by Judge Raulston when the exchanges between Darrow and Bryan became so heated that it was more than apparent that nothing further could be gained for purposes of the trial.

In closing arguments, the defense counsel pleaded for the jury to find Mr. Scopes guilty. The jury returned the desired verdict and then deferred to Judge Raulston to determine the appropriate fine, which he set at $100. This verdict was appealed to the Tennessee Supreme Court in 1927 by the ACLU in their continuing attempt to have the statute declared unconstitutional. Fortunately for the Tennessee state justices, who were trying to avoid further embarrassment for the state over its "monkey law," it was observed that, according to the state constitution, the jury determines any fine of $100 or more rather than the judge. In this case, the jury had deferred to the judge to impose the $100 minimum fine. The state supreme court justices declared a mistrial over this point and threw out the verdict against Scopes. The court then suggested to the state attorney general that he drop the whole matter, which he did, thus foiling the ACLU's intent to appeal to the US Supreme Court.[8]

The overall outcome of the Scopes trial as perceived in the public mind was that religion in general and belief in supernatural creation in particular were ridiculed as outdated superstition. Yet this legacy was instilled more through subsequent historical writing (or more to the point, not so historical misrepresentations) looking back on the decade of the 1920s.[9] The outcome as weighed in the media around the time of the trial was perhaps less decisive. Yet a sharp bias against belief in the Bible was evident in many news reports as well as histories. The Darrow-Bryan interrogatory was the chief catalyst for the Scopes trial legacy, as it had made a serious impression on the public conscience about the validity of the Bible and the Genesis record, and reporters and later writers were able to exploit that. Consequently, since the 1930s, fundamentalist beliefs have often been construed as anti-intellectual and unworthy of consideration in matters of public debate. Yet because of the effect on the textbook market from the restrictions on teaching evolution in the South, science textbook suppliers removed nearly all of the evolutionary content until the 1960s.[10] And because of the absence of evolutionary content in textbooks, teaching of evolutionary science disappeared in public schools throughout the United States. The Butler Act was not repealed by the Tennessee Legislature until 1967.

Beneath the machinations and drama of the summer of 1925, there lurked ideological conflicts involving academic freedom, majority rights, the meaning of life, and the proper domains of science and religion. These questions were not and could never be resolved by court trial, by press, or by popular opinion. They continue to haunt the nation eighty-five years later.

Reenter Evolution

Discoveries and advances in genetics from the 1920s through the 1940s and the discovery of DNA in the 1950s resulted in what is known as the neo-Darwinian synthesis. Darwin's natural selection was joined to the empirically verified concept of gene mutation, and the combination was declared the driving force of evolutionary development in living things. With this synthesis, evolutionary theory, which had seen no further breakthroughs since Darwin, revived and was ordained the unifying concept of biology and thus a fully integral component of modern science. The achievement of the neo-Darwinian synthesis greatly increased confidence among scientists that they had at last mastered the subject of biological origins. But as natural selection returned with the neo-Darwinian synthesis, so did its logical offspring. With its random and accidental variations, many scientists and philosophers concluded on the basis of the synthesis that the purpose of existence is superfluous at best or that there is no purpose to life at all.

Thus, a certain shift in the philosophy undergirding culture was imminent. Through the middle decades of the twentieth century, as the role of Christianity in society continued to be diminished by liberalization from within and increasingly marginalized by the intellectual elite without, the atheistic ideology described in chapter 4 known as secular humanism became a fashionable alternative as a basis for ethics, morality, and law. Humanists and social progressives were rolling out the premise that goodness is an inherent attribute of humanity and that humanity can achieve self-fulfillment and ethical conduct apart from God. The successes of evolutionary science generated confidence among progressives and fit perfectly as the justification for this type of belief system. This was the cultural trend wherein evolution returned to public schools.

In line with this trend, another contributing factor was a 1955 play loosely based on the Scopes trial titled *Inherit the Wind*. *Inherit the Wind* harshly stereotyped conservative Christian faith as oppressive, intolerant, and hypocritical; proponents of free thought and evolution were portrayed as unfairly censored. Popular arts and entertainment often build on and even generate a consciousness of history in the public mind, and this play was no exception. The legacy of and perceptions of the Scopes trial have been seriously affected by this play—that is, people tend to understand the significance and implications of the Scopes trial more from the 1955 play than from the 1925 trial itself.[11] It was made into a movie in 1960 that enabled it to reach a wide audience.

Then suddenly, reentry of evolution into public school biology courses was in high gear. Through the 1950s, with the Cold War raging, the nation was overtaken by anxiety that the Soviets were surpassing the United States in scientific progress and science education. The occasion of the successful launch of a Soviet satellite in October 1957 provoked an urgent call to bring science education up to date with modern developments because it was observed that science textbooks in use at that time had changed little over the past several decades. Through a major initiative involving scientists and textbook writers and with ample funding from the National Science Foundation, science textbooks and science teaching were quickly brought up to date. So, except in states where it was still prohibited, evolutionary science returned to science curriculums of public high schools in the early 1960s.

CREATION SCIENCE

After the 1920s, the reasons for public activism against evolution had waned. Evolution was out of the schools. But now, with a newer and stronger version of Darwinism, the most atheistic strain of evolutionary thought, back in the schools, the controversy began to simmer and boil over a third time. Again, what was being taught in the schools was the source of conflict. A significant minority of citizens regarded evolution as subversive to religious faith and opposed its presence in the public school curriculum. Yet another louder minority demanded that evolution be taught as an integral part of modern science.

In 1961, John C. Whitcomb, a professor of theology, and Henry M. Morris, a professor of civil engineering, published *The Genesis Flood*. This book detailed a reexamination of the state of knowledge of geology, showing how the uniformitarian approach could be replaced by a worldwide catastrophic flood as the explanation of rock formations. The book approached the subject from a scientific perspective with the flood of Genesis 6 through 8 in mind. (See also chapter 3 for a discussion on the merits of this approach.) The flood approach was not new as it dated back to the catastrophist school at the beginning of geologic science in the late eighteenth century. In the 1830s, when uniformitarian geology displaced catastrophism, flood explanations were altogether dropped. Then again, in 1923, an attempt to reconcile geological facts with the Genesis flood was revived by George McCready Price, an amateur geologist, in *The New Geology*. The idea was dismissed in the 1920s even by creationists, but Whitcomb and Morris revived it now again with a degree of success. This was the beginning of a new creationist initiative known as creation science, which included a return to recent creation advocacy consistent with the traditional Genesis interpretation.

As an initiative to bring scientific credibility to the Genesis record of creation and the flood, creation science was the strategy undertaken by the Creation Research Society (CRS), which was established in 1963 by Morris and others. The goal was to take the facts of biology and geology and reinterpret them to correlate with the six-day creation and worldwide flood. The appearance of age hypothesis originally suggested by Phillip Gosse in the late nineteenth century was adopted to argue against the age of the universe as determined by developments in astronomy and physics. It was thought that scientists and education professionals would be convinced of the validity of creation science and thereby remove evolution from public education or at least have creation science presented as an alternate theory with equal merit.

A new organization called the Institute for Creation Research (ICR) was subsequently established in 1972 by Henry Morris and has continued to the present day. The tactics were changed to stress public education and outreach rather than targeting professionals in science and education as CRS had done, but still with the chief objective to establish the credibility of creation science. Through various publications and educational programs for churches and schools and lectures at universities, the case for creation science continued to be argued. In 1974, Morris's book, *Scientific Creationism*, supplied a comprehensive foundation for creation science. The creation science initiative gained a significant following within the conservative wings of Protestant denominations but failed to convince those who were not already predisposed to accept it on biblical grounds. Even if their influence outside of conservative Protestantism was minimal, Henry Morris and the ICR may be credited with a triumph of intellectual honesty, putting the unadulterated interpretation of Genesis 1 back on the map.

Meanwhile, there were developments in the courts. In 1965, the Arkansas Education Association, an affiliate of the National Education Association, undertook to challenge the constitutionality of Arkansas's Scopes era antievolution statute in a case known as *Epperson v. Arkansas*. The ACLU subsequently joined in the challenge to this statute. The nominal plaintiff was a newly hired tenth-grade biology teacher purportedly worried about possible prosecution in connection with using a school's administration-approved textbook containing a chapter on evolution. After a lower court struck down the 1928 statute, the Arkansas Supreme Court, in a 1966 two-sentence pronouncement,[12] sustained the right of the state legislature to regulate the teaching content in public schools of the state. The court also called attention to an interesting question on whether the aged statute prohibited any teaching of evolutionary theory at all or only prohibited teaching evolution as being true. Yet regardless of what the original intent was, the law presented an unacceptable obstacle to progress in science education; the question was moot.

The Arkansas Supreme Court's decision was appealed to the US Supreme Court and overturned in 1968. The US Supreme Court ruled that the antievolution law had a religious purpose in banning one particular scientific subject area, biological evolution, because it conflicted with a particular religious doctrine. "There is and can be no doubt," said the court, "that the First Amendment does not permit the state to require that teaching and learning must be tailored to the principles or prohibitions of any religious sect or dogma." With this, the Arkansas statute was declared an unconstitutional establishment of religion in violation of the First and

Fourteenth Amendments to the US Constitution. The religious purpose could not be linked to the wording of the law itself but rather to its association with fundamentalist advocacy in the 1920s. Though rulings of the courts, including the US Supreme Court, apply only to the case under review, all courts rely extensively on previous decisions as legal precedence, and so practically speaking, the right of any state to prohibit the teaching of evolution was ended.[13] The ACLU had finally achieved its Scopes trial objective forty years later. The Tennessee legislature, observing the trajectory of the Arkansas case's proceedings during 1965 to 1968, and under pressure from the ACLU and the National Science Teachers Association, repealed its 1925 antievolution law.[14]

Unable to keep evolution from being taught, creation science advocates launched a political initiative in the late 1970s to require, through legislation, that equal-time or balanced emphasis be given to creation science. The goal was to permit the competing theories of evolution and creation science to be taught in science classes. Students would then be free to decide between the two. Although the ICR's strategy continued to be to sway public opinion through educational programs rather than impose a legal solution on public opinion, legislation seemed the more expedient approach, and it prevailed in the enthusiasm to rectify the unequal situation in science classrooms. Equal-time legislation was proposed in twenty-seven states, becoming law in Arkansas and Louisiana in 1981.

In Arkansas, the equal-time law for evolution and creation science was passed by the state legislature and signed by the governor in 1981. It was quickly challenged in federal district court in the case *McLean v. Arkansas Board of Education*. The challenge was again brought by the ACLU with assistance from a large New York law firm representing a diverse collection of plaintiffs: assorted Protestant denominational bishops and other clergy, Roman Catholic bishops, Jewish organizations, state and national educational organizations, and individual parents. The case centered on whether creation science was really science or religion in disguise. Federal District Judge William Overton adopted a definition of science offered in testimony by evolution historian and philosopher of science Michael Ruse (a Canadian) consisting of five criteria summarized as follows:

1. It is guided by natural law.
2. It has to be explanatory by reference to natural law.
3. It is testable against the empirical world.

4. Its conclusions are tentative.
5. It is falsifiable. (It may potentially be proven false.)

Creation science was ruled as failing to conform to any of these requirements. Having also researched the legislative history of the Arkansas statute and upon an examination of the text of the statute itself, a religious motivation to insert a literal interpretation of the book of Genesis into the state's public schools was concluded to have existed. Consequently, the Arkansas statute requiring that creation science be taught if evolution were taught was ruled as promoting a religious teaching in its purpose and in its effects. If allowed to stand, it would have further, noted Judge Overton, caused excessive entanglement of state government in matters of religion because of the need for "delicate religious judgments" and constant monitoring of classroom materials and discussion to ensure that religious references were avoided. These arguments of Judge Overton followed criteria previously established in the 1971 US Supreme Court case *Lemon v. Kurtzman*.[15] The statute was thus ruled unconstitutional under the Establishment Clause of the First Amendment. Arkansas declined to appeal its case to the US Supreme Court.

As a side note, evolutionary science would not have satisfied any of these five criteria either. Evolutionary theory fails to conform to these criteria because of certain assumptions it needs to make about the unobserved past. Though evolutionary science appeals to natural law, it cannot be said to be guided by natural law or explanatory with reference to it when evolutionary developments have been neither observed in nature nor repeated in a laboratory. Evolutionary theory is not testable as far as it concerns itself with unique events that happened in the past. The assertion that evolution has occurred has never been tentative because it functions as a premise, not a conclusion. Nor can it be falsified because that is science's only opportunity to explain how living things arrived here. But ruling whether evolution is science or religion in disguise was not within the scope of this case.

The similar equal-time, balanced treatment measure instituted by Louisiana state government in 1981 was likewise challenged by the ACLU in the US Supreme Court case *Edwards v. Aguillard*. After the case was tried in Louisiana Federal District Court, where it was overturned in 1985, the verdict was appealed by the state to the US Supreme Court, which ruled on the case in 1987. The court upheld the lower court's decision ruling Louisiana's law unconstitutional, applying the same criteria as in the *McLean* case from *Lemon v. Kurtzman*. The court also found the equal-time law to be one-sided in its preferential treatment of creation science. It did not give teachers a flexibility that they did not already have to teach other theories besides evolution but rather only placed restrictions against teaching evolution if creation science was not also taught. As such, it did not fulfill the stated purpose of advancing academic freedom. Certain statements from the court's ruling are noteworthy:

> 1. (b) The Act impermissibly endorses religion by advancing the religious belief that a supernatural being created humankind. The legislative history demonstrates that the term "creation science," as contemplated by the state legislature, embraces this religious teaching. The Act's primary purpose was to change the public school science curriculum to provide persuasive advantage to a particular religious doctrine that rejects the factual basis of evolution in its entirety. Thus, the Act is designed either to promote the theory of creation science that embodies a particular religious tenet or to prohibit the teaching of a scientific theory disfavored by certain religious sects. In either case, the Act violates the First Amendment.

> We do not imply that a legislature could never require that scientific critiques of prevailing scientific theories be taught. Indeed, the Court acknowledged in Stone [*Stone v. Graham*] that its decisions forbidding the posting of the Ten Commandments did not mean that no use could ever be made of the Ten Commandments, or that the Ten Commandments played an exclusively religious role in the history of Western Civilization. In a similar way, teaching a variety of scientific theories about the origins of humankind to schoolchildren might be validly done with the clear secular intent of enhancing the effectiveness of science instruction. But because the primary purpose of the Creationism Act is to endorse a particular religious doctrine, the Act furthers religion in violation of the Establishment Clause (*Edwards v. Aguillard*, 482 U.S. 578 (1987)).

In view of the latter excerpt, what might constitute an acceptable critique of evolution or an acceptable alternate scientific theory of origins remains an open question.[16]

The arguments of creation science had failed to convince more than a few scientists or gain a foothold in public schools, and after *Edwards v. Aguillard*, equal time for creation science was no longer a legal option. Even more than in 1925, these court cases were attended by subtle and sometimes not-so-subtle ridicule of creation and the Genesis record by an opportunistic mainstream media while they were in progress: "How could anyone be so credulous as to take seriously the Genesis creation 'story?'"

Following these court defeats, the creation science movement became somewhat unglued. Other independent initiatives began to arise. Disagreement on the age of the earth and the universe became a significant factor dividing creationist energy and focus. The ICR's creation science remained aligned with the literal interpretation of Genesis 1 on the age question. Yet many other creationists were standing by an age of the earth and the universe consistent with geology and astronomy, though still rejecting biological evolution. While creation science continued to be an active movement through the ICR's outreach, another strategic approach was emerging.

INTELLIGENT DESIGN

The intelligent design movement is currently the leading edge of alternative biological origin theory in public science education. It was adopted as a strategic focus after *Edwards v. Aguillard* partly in order to satisfy the wording of the second quoted paragraph from the US Supreme Court's ruling on that case above. This strategy incorporated lessons learned from experiences with creation science: purge the use of the term *creation* and all references to Genesis, focus on evidence that evolution cannot explain, and argue for the existence of nonspecific, nonnatural intelligence exclusively from scientific data. These differences in emphasis, it was hoped, would avoid the difficulties encountered by creation science in the courts that recognized it as a religious viewpoint. The approach of intelligent design is to show how the universe, the earth, and living things are not ultimately explainable by natural causes alone and how patterns and complexities of nature indicating design are apparent or at least that they may be inferred. No particular hypothesis as to what intelligent, nonnatural causes might be involved is offered.

Two of the intelligent design movement's chief advocates are biochemistry professor Michael Behe and retired law professor Phillip Johnson. Johnson's 1991 book *Darwin on Trial* was instrumental in exposing the philosophical root of evolutionary science. In *Darwin on Trial*, Johnson showed how evolutionary theory is more a deduction from naturalistic philosophy than a theory having foundations in empirical science. In reviewing the categories of scientific evidence commonly put forward as proof of evolution (such as the fossil record, genetics and mutations, and microevolutionary variability), he concluded that evolutionary theory does not have the empirical foundation the public has been led to believe exists. Small incremental changes over long periods of time adding up to major changes is not what the evidence says happened. Rather, the evidence suggests abrupt and major changes at infrequent intervals. Furthermore, only by extrapolation can the types of microevolutionary changes within species observed today be thought possible to have produced macroevolutionary changes leading to new species in the past. Small incremental changes over long periods of time adding up to major changes is, therefore, a philosophical predetermination, not an inference from empirical evidence. Said Johnson:

> If our philosophy demands that small changes add up to big ones, then the scientific evidence is irrelevant. What Darwinists need to supply is not an arbitrary philosophical principle, but a scientific theory on how macroevolution can occur.[17]

Moreover, said Johnson, the philosophical content of evolutionary theory comes to the public "cloaked in the authority" of empirical science.[18] That is, evolutionary science is made to appear as if it was a product of the scientific method when it is not.

In 1996, Michael Behe published a second milestone book for the intelligent design movement, *Darwin's Black Box: The Biochemical Challenge to Evolution*, in which he argued the case for irreducible complexity.[19] Irreducible complexity is an assertion that a minimum level of complexity and functional specialization is essential to all viable living organisms and that incremental evolutionary steps leading up to the minimum level of complexity could not have comprised viable living organisms. In other words, even the simplest organisms are far too complex to have evolved through numerous incremental stages where various semideveloped parts would have had to exist without yet fulfilling any survival function. It is inconceivable that natural selection could have brought these parts into functional use because they would have had no survival benefits for a population while in intermediate stages of development. The irreducible complexity argument for intelligent design in *Darwin's Black Box* is generally credited with moving intelligent design into the mainstream of the evolution-creation debate.

Opinions on whether irreducible complexity has been addressed adequately by the scientific community are divided along familiar lines. Attempts to explain it away have appealed to a few instances on the subcellular level, where an assembly resembling a subset of components of a supposed irreducably complex structure is seen to perform a certain function in a different species of single-celled organism. But the function the subassembly performs is entirely different than the function of the otherwise irreducably complex structure. To date, scientists have dismissed irreducible complexity as a noncredible argument against evolution; when the idea that natural causes are sufficient to explain all living things is already assumed, irreducible complexity cannot be valid. There is a humorous side note: Behe used a five-part mousetrap to explain the concept of irreducible complexity. Take any one part of the mousetrap away, and the mousetrap will not work. Defiant scientists promptly devised working mousetraps with fewer than five parts.[20]

Besides irreducible complexity, a second argument for intelligent design comes out of recent progress in the science of genetics. The question has arisen as to the origin of the information content of DNA genetic code. Such information, it is contended, can only come from intelligence. Neither operations of the laws of nature nor random processes in nature have ever produced information of any kind (information being generally defined as forms of representation of data, facts, instructions, knowledge, etc.). It cannot be explained by the usual scientific reduction of everything to matter and energy. Information is neither matter nor energy nor a combination of the two but an arrangement of matter or energy. And such an arrangement is not a necessary property of matter or energy. It is further seen that information can be copied or transferred from one medium to another, so it is not dependent on specific matter or specific energy. To date, the response of scientists to this challenge is similar to their response to irreducible complexity: The idea that genetic information can only be attributed to intelligence is not compatible with the naturalistic premise of science, so it cannot be true.

Compared to creation science, intelligent design is a much more generic formulation; it is completely nonspecific as to what intelligent entity(s) might be involved. No position on age is taken either, though generally, ages consistent with conventional geology and astronomy are accepted. Evolutionary biology alone is the object of contention. Where creation science rallies natural causes and the laws of science in support of the book of Genesis's supernatural creation, intelligent design argues only that supernatural causes preceded natural ones. Yet, as generic as it is, it is likewise being considered a religious viewpoint by stewards of public science education.

In 2004, the school board of Dover, Pennsylvania, adopted a disclaimer policy stating that evolution is a theory with gaps in supporting evidence. The disclaimer further stated that while offering an opportunity for students to voluntarily learn about intelligent design, evolution would continue to be required learning to satisfy state academic standards. This policy was challenged in the 2005 federal district court case *Kitzmiller v. Dover Area School District* (Pennsylvania) by the parents of some students and the ACLU, who were joined by Americans United for the Separation of Church and State. Federal District Judge John Jones ruled that the school board's policy was unconstitutional because intelligent design is not science but rather endorses a religious view. Specifically, the reasons for this ruling consisted of the following points:

1. An expressly religious motivation was evident in both the language of the school board in defending the policy and in the historical development of intelligent design.
2. Even mentioning the words *intelligent design* by a teacher in class would generally be perceived by informed observers as a religious endorsement.
3. Intelligent design, in contrast to science, appeals to supernatural explanations and is thus inappropriate in science classes.
4. The logic of using arguments against evolution as grounds for design is flawed.
5. The court perceived that all arguments against evolution had effectively been refuted by scientists.

Judge Jones further stressed that intelligent design "has failed to gain acceptance in the scientific community. It has not generated peer-reviewed publications, nor has it been the subject of testing and research." (*Kitzmiller v. Dover Area School District*, December 20, 2005) Intelligent design thus failed to qualify as science.

During courtroom testimony in the *Kitzmiller* case, biologist Kenneth Miller, an expert witness for the plaintiffs, took the stand with a rather unusual device attached to his tie. It was a mousetrap with some of the parts removed, and he was using it as a tie clasp. This was an example of reducible complexity—a "reduced" mousetrap serving an equally legitimate function. No one seems to have thought to throw the ploy back the other way, that both the mousetrap and the "tie clasp" were examples of intelligent design. The question then becomes: If intelligent designs exist, and as we are assured by scientists that they do not exist in nature and thus do not arise from nature, then where do they come from?

Intelligent design has been the best contender to date with any possibility of rectifying the absence of a "variety of scientific theories" in public school biology education. The evidence being presented is a formidable challenge to the scientific community's resistance to any theory involving nonnatural causes. Nevertheless, it appears unlikely at this point that any foot in the door to public science education can be gained. Even if it were successful, another source of controversy might arise. There would be no scientific means to fill the vacuum as to what might be known about said intelligence. Any questions about the intelligent designer are unanswerable. This is an obvious difficulty in arguing the existence of the supernatural exclusively from the natural world. One can argue for its existence but not more than that.

OTHER INITIATIVES

Complementing intelligent design and creation science, other initiatives have repeatedly been put forward to contend against the teaching of biological evolution in public-school science education. These initiatives have not offered any replacement theory but have either (1) required evidence against evolution to be taught or (2) required disclaimers on biology textbooks to identify evolution as "just a theory." Neither of these approaches has had permanent success to date in blunting the ideological edge of evolutionary science education. The following list is a representative though not comprehensive list of such initiatives over the past few decades. Most have either failed or have been short-lived. Many more similar initiatives have been proposed at state and local levels throughout the nation, but they have never achieved state legislature or school board approval.

- Texas 1970–84, State Board of Education biology textbook disclaimer stating that evolution is a controversial theory and not a fact.

- Tennessee 1973–5, legislation requiring balanced emphasis for creation science and disclaimer that evolution is only a theory, not a scientific fact.
- California 1969–?, State Board of Education "antidogmatism" policy that requires all scientific evidence be considered.
- California 1993–5, Vista School Board: a brief and controversial balanced treatment initiative consistent with California's antidogmatism policy.
- Louisiana 1994–7, Tangipahoa Parish Board of Education "evolution is a theory" disclaimer (overturned in federal court).
- Alabama 1995, short-lived State Board of Education biology textbook disclaimer stating that evolution is a controversial theory and not a fact.
- Tennessee 1996, legislation proposed to oppose teaching evolution as fact (failed to pass).
- Ohio 1996, legislation proposed to require teaching evidence against evolution (failed to pass).
- Washington 1998, State Board of Education disclaimer similar to that of Alabama (failed to pass).
- Kansas 1999–2001, State Board of Education policy removing references to macroevolution, age of the earth, and big bang theory from state education standards and assessment testing.
- Ohio 2002–6, State Board of Education policy to allow evidence against evolution to be taught.
- Oklahoma 2003, State Board of Education disclaimer similar to that of Alabama (failed to pass).
- Mississippi 2003, State Board of Education disclaimer similar to that of Alabama (failed to pass).
- Georgia 2002–5, Cobb County School Board "evolution is a theory" disclaimer (overturned in Federal Court).
- Kansas 2005–?, State Board of Education policy to allow evidence against evolution to be taught.

What to Make of This

Why has creation been so marginalized in science education? Why have no alternate scientific theories besides evolution succeeded in gaining entry into science education as the US Supreme Court suggests may be allowed? And why has it been so difficult, even impossible to restrict or even criticize evolutionary science education?

Since the 1960s, whenever the controversy has found its way into the courts, it has arrived as a case of an infringement of rights under the First Amendment to the US Constitution. Through prior legal precedent in interpretation, the First Amendment's Establishment Clause has come to mean that no government authority, whether state legislature or local school board, may endorse, get entangled in, or promote religious beliefs. Evolutionary theory is perceived as science. It has all the appearances of science and the practice of the scientific method behind it. It does not have the look, sound, or feel of a religion. Therefore, it does not come under suspicion as religion. Conversely, creation science states and intelligent design implies that God exists. In referencing the book of Genesis or even the most generic and nondescript idea of a nonnatural entity, creation science and intelligent design are inevitably perceived as religious beliefs. Scientists, education professionals, and many parents cry foul wherever such ideas dare to defile the science classroom.

Attempts to argue that evolutionary theory is religion in disguise (or more realistically, an ideology upon which religion such as secular humanism is built) or that creation science and intelligent design have scientific credibility have not fared well. Notwithstanding the merits of these arguments, perceptions and biases to the contrary are strong and are cultivated by the educational system itself and by the media. The ACLU, the National Center for Science Education (NCSE), and various other organizations, having assumed self-appointed roles as separation-of-church-and-state police, thus have the weight of constitutional law in their favor.[21]

Compounding the difficulty are many who are only too happy to accept both God and evolution and who say there exists no conflict between them, that evolution in no way precludes the existence of a supernatural creator. The idea that God used evolution as the means of creation makes sense to them. As long as you do not look for evidence that the proposition may be true, theistic evolution is an appealing compromise. All you need to do is make a few slight adjustments in interpreting the book of Genesis ... or ignore it completely.[22] Many wonder what all the fuss is over and why fundamentalist-oriented religion cannot simply keep to its own domain of influence and stay out of science. It is noteworthy that religious denominations that formerly turned their back on the honest interpretation of Genesis because they could not square it with the progress of science were seen as plaintiffs in *McLean v. Arkansas*. The courts have thus considered that evolution is not hostile to religion *per se* but only in particular. This makes it impossible to craft a statute that does not appear to favor religion in particular and thereby appear as an establishment of religion.

Where does this leave creation in the science classrooms of public schools? Are there other arguments that can be made or other strategies? If a scientific theory is what is being opposed, that would call for further scientific counterarguments and evidence. But if it is understood that what is opposed is more an ideology disguised as science than a scientific theory, other strategies may be in order.

It is advisable to take a moment to clear up confusion about there being a multiplicity of alternative views as implied in some of the court opinions. The practice of modern science is predicated on the exclusion of nonnatural causes. This practice is traceable back to the scientific revolution as we have seen in previous chapters. In order to assure science's claim over the subject area of origin and keep competing ideas out, exclusive adherence to natural causes is enforced through peer review. The only plausible way to do this is by maintaining the idea of small, incremental steps spaced out over long spans of time, presupposing that numerous small steps add up to big changes. That is, of course, what the term *evolution* means. So while there are sub-theories within evolutionary science whose merits are debated,[23] evolution itself is not debated, nor can it be debated without risk of invalidating science's claim over origins.

Even intelligent design, which from beginning to end maintains an exclusive focus on what may be inferred from the natural world, is still a supernatural proposition and impermissible as science. This, again, is why it has been impossible to get anything through the court system that would serve as a counterperspective to evolution and still qualify as science. There are no alternatives besides natural and supernatural. It should be apparent that the absence of other alternatives is more a consequence of a law of logic than a lack of other hypotheses having been proposed. The charge that advocates of creation science and intelligent design pose a false dilemma in basing their claims on a critique of evolutionary theory should not stand until someone proposes a third alternative that is neither evolution nor creation.[24]

So the better question might be why science education is the center of conflict. While everyone seems to accept the evolution-creation debate as a scientific matter, is that where this debate should be or even can be settled? Should the creation science or intelligent design propositions be introduced into other subject areas of public education such as philosophy? The answer is yes, they should be introduced into other subject areas because of the self-imposed limitation of science to natural causes only. But practically, the answer is no. Attempts to permit science's claims over origins to come under suspicion within subject areas outside of science would be opposed just as vigorously by the scientific community and public advocates of evolutionary science. It will be found that separation of church and state works equally well in or out of science classrooms.

CONCLUDING REMARKS

Science has claimed the origin of the universe, the earth, and all living things as its proper domain. The scientific origin theories imply a sufficiency that renders the supernatural as irrelevant, even nonexistent. But the positions of scientists on this point are not entirely consistent; they also admit that nothing can be said by science about anything other than what is natural. While the nonnatural may or may not exist, a determination on that point cannot be made by science. Eugenie C. Scott, director of the National Center for Science Education (NCSE), active in defending and promoting evolutionary science education, acknowledges that ultimate origin is not a matter accessible to science: "We can put on our philosophers' hats and comment as individuals, but as scientists, we can't deal with ultimate cause."[25] The US National Academy of Sciences (NAS) has issued official statements affirming this view: "Science is limited to explaining the natural world through natural causes. Science can say nothing about the supernatural. Whether God exists or not is a question about which science is neutral."[26]

By proposing a partitioning of scientific and religious thought, the academy also implicitly acknowledges the inability of science to explain everything: "Religion and science are separate and mutually exclusive realms of human thought whose presentation in the same context leads to misunderstanding of both scientific theory and religious belief."[27]

Matters of value and ethics, for example, are not matters that can be addressed by science; they are assigned to religion and philosophy.[28] In view of such statements, why then is there such an emphasis on the scientific and on the natural when the subject of origin comes up? There is not an awful lot of consistency surrounding the academy's points because on many occasions, scientists either state or imply that all existence is material and that the origin of everything is science's proper territory. An underlying confusion exists over what science can and cannot determine. Is all existence natural because that is all science can deal with, or has science improperly claimed the subject area of origin?

One thing is apparent: Action or intervention by God can never be a conclusion of science, not as science has been defined over the last century and a half. *"Natural causes only" is a premise; it is not a conclusion of the scientific method as the public has been led to believe.* While this premise may be necessary for progress toward more scientific knowledge, it may simultaneously constrain progress toward overall knowledge. The exclusion of supernatural causes is necessary to do scientific work, but nonscientists are not obligated to accept it.

Now it will certainly be said that creationists are just as biased toward the supernatural as scientists are toward the natural. This may be true. It may be that creationists are biased toward an unprovable claim. (Whether this is provable or unprovable should become evident in the subsequent chapters.) But the scientific bias is just as unprovable! And while science claims to have found answers, they fail as satisfying explanations of the origin of human existence. Hardly anyone can deny that there is more to our existence than matter and energy. With its total naturalistic focus, reducing everything to matter and energy, could it be that science has attempted to chart waters too deep for it? Science's claim that existence is such as it describes is like saying that music and literature are nothing more than rules of harmony and rules of grammar.[29] Scientific accounts of music and literature would entirely miss the point. Failure to come up with satisfying explanations would be enough of a reason to seek explanations elsewhere. If scientists back off from their forceful insistence that evolutionary theory is the answer, no longer would it be necessary to interject supernatural causes, such as intelligent design, into science education.

To summarize the root of the controversy:

1. The general concept of evolution provided science with new avenues of inquiry, inquiry into the past.
2. Science then effectively claimed jurisdiction over matters of origin, displacing theology.
3. Admittedly, the existence or nonexistence of the supernatural cannot be established by scientific inquiry.
4. Science cannot then be certain that it has jurisdiction over all aspects of existence.
5. Therefore, not knowing it can cover all aspects, science cannot be certain about matters involving origin.
6. At any rate, the explanations that science confines itself to, reducing everything to material causes, are unsatisfying.
7. Yet scientists consistently and vehemently resist any interjection of supernatural causes into science.

So then, either science vacates the subject area of origins, or pressure for supernatural intrusions into science education will probably continue.

Just as scientific evidence fails to resolve the controversy, legislation and court rulings on science education likewise fail even more to settle it. The ideological nature of this conflict runs too deep and wide for such superficial solutions. It should be clear now that this controversy has overflowed scientific turf. Judge Overton stated that himself in considering the extent to which evolutionary theory would need to be balanced with creation science in the public schools:

The Act [referring to Arkansas's balanced treatment act, Act 590] prohibits instruction in any religious doctrine or references to religious writings. The Act is self-contradictory and compliance is impossible unless the public schools elect to forego significant portions of subjects such as biology, world history, geology, zoology, botany, psychology, anthropology, sociology, philosophy, physics and chemistry. Presently, the concepts of evolutionary theory as described in 4(b) [of the Act] permeate the public textbooks. (*McLean v. Arkansas*, January 5, 1982, IV(E))

According to Judge Overton, the concepts of evolution theory are not limited to biology or science textbooks. The scope of the controversy is such that we are dealing with something more comprehensive. Before continuing to deal with evolution as an ideology, now more properly termed *evolutionism*, there is a need to bring up something else—something that science depends on but will never be able to fit within its theories of existence.

Chapter 8:
Thought

*The mind that conceives the thought "There
are no miracles" is itself a miracle.*

—Anonymous

The evolutionary view of living things links inherited characteristics to survival. The theory of natural selection says that traits that help organisms to survive are the traits that pass on to the next generation. Among inherited characteristics are patterns of behavior. Thus, behavioral traits that increase the prospects for survival are favored by natural selection over behavioral traits that decrease prospects for survival. Continuing this logic along evolutionary lines, among patterns of behavior that may be included here are patterns of thought, the thoughts of human beings, say. From various thoughts originate various responses to the environment that each have a bearing on survival, whether positive or negative.

The odd thing about this point is that, and taking Galileo as an example, Galileo's behavior at the time of his conflict with the church did not increase his prospects for survival. But now years later, the thoughts and consequent behaviors that got him into trouble are commended. The reason his ideas are now commended has to do with whether or not they were found to be true and nothing at all to do with survival. Thus, there is some ground for suspicion about what is claimed to be the originating factor behind at least some of the activity that falls under the category labeled thought. It appears that natural selection may not be the originating factor. Or else there is an implicit claim lurking somewhere within evolutionary theory about thought, *if naturally selected, then true,* the validity of which is not apparent.

We can proceed further in this direction by noting the existence of contrary propositions and ideas. Leaving aside the question over the propriety of the church's methods of persuasion, why not just let survival mediate between conflicting ideas and see which ones are naturally selected? It is a wonder that anyone would argue over contradictory ideas and not just let the survival test mechanism we call the environment weed out those that would be found deficient in that respect. But people do argue and do attempt to persuade. There is no question that at least some thought is directed toward discerning truth and has nothing whatsoever to do with survival. As science strives for objectivity in its conclusions in the expectation of finding out truth about the natural world, let us define a term that distinguishes that mode of thought: "objective thought," the mental exercise of discerning truth.

The point of this chapter is to distinguish between the thought process that scientists use in drawing objective conclusions from the study of nature and the cause-and-effect processes in nature that are the object of scientific inquiry. Once having gained this distinction, it becomes evident why the scientific thought process is not one of the natural processes that scientists study; it is not a natural process. Neither is scientific knowledge naturally selected as an advantage for survival. (What has been the survival advantage of knowing that the earth revolves instead of the Sun, that the speed of light is always 186,000 miles per second in empty space, or that radio-isotope decay is exponential as opposed to linear?) Of course, objective thought and the drawing of conclusions as well as the existence of cause-and-effect sequences apply equally to other disciplines outside of science, but science will do for the purpose.

A Pivotal Distinction

We all know about cause and effect and use that knowledge all the time in drawing inferences. If we say, "A south wind must have been blowing earlier today because it is hot," we would be inferring the cause of an observed effect, a cause, for whatever reason, not experienced. If we say, "It will be hot today because a south wind is blowing," we would be inferring a future effect from an observed cause. If, on the other hand, we said, "It is hot today because a south wind was blowing earlier," we would be describing a cause-effect relationship where both cause and effect are already known. In the first two cases, there is a venture to obtain new knowledge; in the third, there is no new knowledge but merely a description. In the first two, there are logical relationships between observations and conclusions; in the third, a cause-effect relationship between two conditions. The inferences drawn in the first two cases are based on knowledge of the particular cause-effect relationship described in the third, but the inferences do not spring out of the cause-effect relationship itself. The inferences depend on knowing about the cause-effect sequence, but they are not caused by the observed conditions. Therein lies the first hint of a pivotal distinction. Though the term *because* is common to all three sentences, that should not lead to confusion. One use of *because* introduces logical grounds for a conclusion; the other links an effect to its cause.

We should further consider in practical terms just how the objective thought process, which consists of the drawing of inferences or conclusions, is differentiated from cause-effect processes of nature. This differentiation may be understood by the following observations:

1. Objective thought is *about* something other than itself.

The *aboutness* is evident in any proposition, assertion, claim of truth, or belief *about* something, even about thought itself. Conversely, effects produced by causes in nature are never *about* anything. If we propose that gravity is a force acting upon objects that have mass, we are making a claim *about* gravity that is distinct from gravity itself and distinct from an effect of gravity, such as the falling of an object. If we assert that southerly breezes bring in hot weather, we are making a claim *about* a certain cause-effect linkage that we believe operates in the natural world. But that claim is not the cause-effect linkage itself; the cause-effect linkage, if it exists, exists apart from any belief *about* it.

2. Objective thought may be true, or it may be false.

The very nature of propositions, assertions, beliefs, conclusions, and claims is the possibility of their being true or false. Even if a proposition turns out to be false, that recognition would itself be a truth, and false propositions may be corrected by better reasoning or more reliable premises. But to say that an effect produced by a cause is true or false is meaningless. Effects are inevitable and not subject to deliberation as true or false in the way a proposition is. If gravity were to have caused (in the cause-effect sense) our assertion about gravity, it would be an inevitable effect of gravity and not a proposition that could potentially be true or potentially false. It would not really qualify as an assertion or proposition in that case.

3. Objective thought is deliberate, not inevitable.

Conclusions of scientific inquiry are not inevitable but are drawn deliberately in conjunction with a review of relevant scientific findings and principles. The law of gravity was a result of such deliberation, and the law of gravity came along after gravity had been in existence for some time. It did not need to come along; it was not an inevitable or necessary result of gravity's existence. While some inferences, such as predicting that a south wind will cause hot weather, may be automatically drawn, they are not inevitable and do not need to be drawn. On the other hand, effects proceed inevitably from their causes. The falling of an object is an effect of gravity, and the effect is inevitable once the object has been released in midair.

4. Objective thought reaches beyond experience.

The characteristic of objective thought that perhaps most sets it apart from cause-effect sequences is how knowledge beyond experience can be reached. The scientific knowledge of gravity is well beyond anyone's experience, and in fact, required a great deal of insight into observed effects and an understanding of the nature of space and time in order to arrive at it. Conversely, the most that a cause-effect account of thought can offer is that of responses to sensations, responses well-conditioned perhaps for survival, yes, but not ventures toward insight beyond experience. While hot weather might trigger a response of finding cool shade, insight into the cause of that weather cannot be included among the inevitable effects of experiencing that weather.

It might be said here that an accepted idea, such as that southerly wind brings in hot weather, is the result of accumulated experiences, that the one condition always follows the other. It is true that past experiences produce expectations, but expectation is not the same as inference. It is believed that a south wind produces hot weather, yes. But that is believed not because of a mere pattern of past events and conditions but because, having discerned a pattern, we have gone on to recognize and understand the principles and factors involved in bringing about the pattern. The idea that there is a linkage between southerly breezes and high temperatures is grounded on insights into the principles and factors involved; it is not merely an expectation based on past circumstantial association. In the same way, neither are scientific conclusions mere expectations based on past experience but insights into principles and factors underlying observed patterns in nature.

A UNIVERSAL RATIONALITY

As nature is subject to cause-effect laws discernible from scientific inquiry, so objective thought is subject to certain laws, laws of logic or laws of reasoning. The law of noncontradiction is one of the most fundamental of these laws. In one form, it says that a proposition cannot both be true and self-contradictory.[1] A good example of a self-contradictory proposition is this: *Truth does not exist.* This proposition is put forward as a truth about truth, but it undercuts its own ground in saying that truth does not exist. Therefore, the proposition is not true. What this means is that truth *does* exist, which is a self-consistent proposition.

The law of noncontradiction is a self-evident principle (or axiom) neither admitting nor requiring argument. Truth exists. Everyone is subject to both the law and whatever truth there happens to be. No one can disregard this point if they want to be credible in their assertions. In their self-evidentness, the law of noncontradiction and the existence of truth are not subject to causes or effects in nature either. Even if they were, we would maintain an illusion that it was not so because to think otherwise would mean giving up rational conversation. Indeed, we already assume that it is nature that is the subject. It is assumed that nature is subject to a certain rational ordering and to an ordering not terribly different than laws of reasoning. It is not scientific inquiry that tells us this. Indeed, scientific inquiry depends on this.

It is the sense of rational order that tells us, for example, that laws of nature, as repeatable and predictable cause-and-effect sequences, are universally true and reliable. Laws of nature cannot be understood on experience alone because experience is limited. Again, taking gravity, the effects of gravitation may be observed in the current moment. We may recollect that gravitation has, in the past, been consistent with what is now observed. But what would ever cause us to think that it will continue this way five minutes from now unless we read a sense of order into the universe? Experience does not tell us of future continuity because experience only tells us the present and past. Experience does not tell us that gravity will act the same tomorrow as it has today, but a rational ordering to which nature is subject does tell us that. The insight by which laws of nature are known—or indeed knowledge of anything beyond what is experienced—can only be attributed to a rationality to which nature is subject and which the mind can grasp and leverage. It is this rationality on which the validity of inference and knowledge beyond experience rests.

Thus, all that is known beyond what can be experienced directly is inferred from experience on the belief that the world is rational. Without inference and belief in a universal rationality, nothing can be known beyond the sensations that come to us through the five senses. The sensations we experience do not just bubble up at random; we accept that there is a logic both to them and behind them that can be leveraged to figure out things beyond them. "It is clear that everything we know, beyond our own immediate sensations, is inferred from those sensations," wrote C. S. Lewis, "for we can know nothing beyond our own sensations at the moment unless the act of inference is the real insight that it claims to be."[2] And the act of inference depends on belief in a universal rationality.

An Attempt at Natural Cause

Perhaps there is some way to cast a cause-and-effect account for objective thought. The idea that the mind comprehends laws of nature because laws of nature produced the mind might be considered a satisfactory explanation of how natural cause and effect produced rational thinking. But that would mean that any product of laws of nature might comprehend laws of nature. How is the mind to be distinguished from other things?

Notwithstanding how that distinction might be made, rather than assume the point of contention, that laws of nature produced the mind, the point should be justified somehow. It is already known that the mind comprehends laws of nature. What is not known and the question to be answered is this: What produced the mind? Thus, the assertion should be rephrased: Laws of nature must have produced the mind because the mind comprehends them. The cause and effect *because* is thereby replaced with the *because* of rational inference. The idea is now not so attractive as it first appeared; the conclusion does not necessarily follow. (There could still be minds that comprehend laws of nature that were not produced by laws of nature.) And if the laws of nature that supposedly produced the mind fall back on the theory of natural selection (and so far, natural selection is the only scientific theory offered that would venture to explain the mind), then we are back to the idea entertained at the beginning of this chapter, that thought evolved against a survival criterion and not a truth criterion.[3] A survival criterion offers us no assurance that any of our thoughts could be true. The proposition "if naturally selected, then true" then falls under the same suspicion as any other.

It appears that the validity of rational inference and objective thinking must be assumed upfront in order to gain insight into anything at all. Once having granted that assumption, a cause-and-effect account, such as that offered by evolutionary science, can add no further validity. No matter, the process of rational inference that produces knowledge, such as the knowledge of the laws of nature, is the thinking that appeals to reasons for it, not causes for it. If knowledge were an effect, as C. S. Lewis noted, it would be "as if cabbages, in addition to resulting *from* the laws of botany, also gave us lectures on that subject."[4] The absurdity of attributing knowledge to natural causes becomes apparent as one attempts to so attribute it.

There is a distinction to be made then between caused and inferred: Sensations are caused; knowledge is inferred. If knowledge really exists, there must be a subset of thought that is solely dependent for its existence on its being true and not caused.[5] That subset is what we are calling objective thought. But the moment a cause-and-effect process has any bearing on our thinking, that thinking becomes subject to conditions and ceases to be objective. We know how thinking may be clouded by momentary pain, euphoria, anxiety, fear, anger, or other emotions traceable to conditions. With objective thinking so intruded upon by conditions, no one could seriously regard their attempts to discern truth as *arising* from conditions, effects, as it were, of them.

FURTHER RAMIFICATIONS

Conveying and receiving the patterns and images of thought by means of language requires a certain degree of objectivity in the meaning of words. Also required are logical constructions that order words into sentences according to objective rules of syntax and grammar. Even with all the subjectivity that exists from one person to the next, how perceptions are subject to that which makes individuals unique, the meaning carried by spoken or written words and sentences must remain (to a large degree) unaffected by participants in a communication. This assertion about objectivity in language becomes obvious the instant one tries to deny it. It is impossible to deny objectivity in language without affirming it at the same time; any denial is advanced as objective. The law of noncontradiction is at work here. And in raising an objection, the contrary assertion is assumed to be a superior one, and objectively so. Otherwise, no one would bother to argue over the truth or falseness of this assertion or any other for that matter. If language is thus a carrier of objectivity, then language cannot be a cause-and-effect process and cannot be an effect of causes in nature. The origin of language must fall alongside that of objective thought.

Where does objective thought stand in relation to probable physical manifestations of it? Do electrochemical interactions involving molecules in the brain produce thinking, or does thinking drive the interactions of molecules? If we choose that interactions produce objective thought, how would the four distinctions listed above be established? Someone needs to come up with a way to explain that. It cannot be done. It is quite true that we do not know what happens to thought or to the mind without a brain. Perhaps the mind disappears. Or perhaps nothing happens to it and it remains the way it was without a brain, though not accessible once the mechanical apparatus of language is removed. It does not seem possible that thinking can be explained either as a total natural process or as a total nonnatural process.

One observation assures a conclusion on the mental-physical question, though in a perhaps unexpected direction. The observation is this: The linkage between thought and its manifestation as language is seamless. There is no awareness of a translation process between thoughts and, say, the words spoken to communicate them, or between speech heard and thoughts in turn generated. The translations from mental to physical and back again proceed automatically. As astonishingly complex as the mental-mechanical translation mechanisms between thought and speech and between hearing and thought must be, one proceeds from the other without the slightest awareness of an intervening translation process. The idea that a scientific account can still be given for brain, tongue, and ear separate from language and purposeful mental activity is dubious. The seamlessness of translations from thought to speech and from speech back to thought testify to a common origin for the speech, hearing, and thought apparatus necessary to facilitate language and language itself.

Concluding Remarks

We are seeing that objective thought cannot be reduced to something arising from nature without invalidating it. If objective thought cannot be explained from within the natural world, the only other possibility is a source outside the natural world. There are some things that science, in principle, cannot and will never explain. Objective thought is one of those things. But it is precisely this kind of thought on which science depends. Science has an explanation for everything it seems, but it cannot and will never explain itself. Writes Lewis:

The Naturalists have been engaged in thinking about Nature. They have not attended to the fact that they were thinking. The moment one attends to this it is obvious that one's own thinking cannot be merely a natural event, and that therefore something other than Nature exists. The Supernatural is not remote and abstruse, it is a matter of daily and hourly experience, as intimate as breathing.[6]

It is a matter of daily experience that rational thoughts induce and enable us to alter the course of Nature. On the other hand, Nature is quite powerless to produce rational thought: not that she never modifies our thinking but that the moment she does so, it ceases (for that very reason) to be rational. For, as we have seen, a train of thought loses all rational credentials as soon as it can be shown to be wholly the result of non-rational causes.[7]

Objectivity in science depends on the validity of rational inference, and rational inference has validity only if a universal rationality that transcends both nature and the mind is assumed. Unless rational inference is accepted as valid on that basis, all knowledge is reduced to sensations of the moment and scientific knowledge that extends beyond those sensations has no foundation. While the assumed validity of rational inference is necessary—and doubtless everyone operates on that basis—should we never ask why that is so?

Chapter 9:
Philosophy 1

*There is nothing so strange and unbelievable that it has
not been said by one philosopher or another.*

—Rene Descartes,
the father of modern philosophy

PHILOSOPHY?

Paul wrote, "See to it that no one makes a prey of you by philosophy and empty deceit, according to human tradition, according to the elemental spirits of the universe, and not according to Christ."[1] One might think of this as a warning to stay at a safe distance from philosophy. The point, however, is that there is sound philosophy and unsound. The sound is "according to Christ"; the unsound is "according to human tradition, according to the elemental spirits of the universe." This is not so much a prohibition against philosophy. It is more a warning to make sure that any philosophy subscribed to is properly grounded, and a counsel to be ready to recognize an improperly grounded philosophy, seeing it for what it is.

The discipline of philosophy has a great deal of experience with some very persistent and seemingly unyielding questions, and it is worthwhile to take a look at it. Here is an assorted list of some general philosophical questions:

- How does one know anything?
- What is the meaning of life?

- What is good, and what is evil?
- Why is there evil?
- What defines a person?
- Is there free will?
- Is there existence after death?
- Does God exist?

Some of these questions are linked to the subject area of origin. In fact, all philosophical questions are in some way linked to that subject. Whether as philosophy or as science, questions about origin are asked. In so asking, a coherency and unity is sought that somehow connects the diversity experienced in the world to some larger purpose or overarching logic. Intuitively, true philosophy should be that which brings the most coherency to existence—that is, the most explanatory power. That intuitive sense was certainly seen to be present in science. Coherency and unity are at least some of the impetus behind philosophical as well as scientific inquiry.

At the personal level, four philosophical questions have been proposed that condense what needs to be answered in order to reach a coherent and relevant as well as practical result:

- Origin: Where did I come from?
- Meaning: Why am I here?
- Morality: How should I live?
- Destiny: Where am I going?[2]

This chapter and the next serve as an introduction to thinking about the subject of origin philosophically. In addition to having a scientific dimension, the evolution-creation controversy also has a philosophical dimension. (In fact, as we have seen, science never truly left its philosophical roots behind.) It is philosophical methods that are of particular interest here. Philosophical methods can help in judging the validity of assumptions and in clarifying implications of opposing views (the opposing views in the question at hand being the polarity between the claim that all things have a natural origin versus the claim that all things have a supernatural origin). Beyond that, there is a need to know something of the anatomy of sound thinking and reasoning, effectively dealing with contrary predispositions, persuading the skeptic. Over the remaining chapters, we will begin to see how foundations other than Christ, particularly what evolutionary science has offered, will not stand up to philosophical inquiry.

TROUBLE

There is trouble in philosophy, and that trouble lies with presuppositions. Presuppositions are what are assumed at the outset of a rational argument or an inquiry; they function as premises. The reason for this trouble is that presuppositions are just that, presupposed; they are assumed. A train of logical inferences may be quite sound but lead to a false conclusion because one or more premises are false. This is the same sort of problem that science can (and has) run into. Some examples of statements that could serve as premises include the following: All men are mortal. All dogs are brown. All objects fall to the center of the universe. All that exists is natural. They can be true or false statements. (For an introductory discussion on validating premises, see Appendix B.)

Premises and presuppositions cannot generally be overthrown by reasoned argument because there are typically other factors besides reasoned arguments for choosing them in the first place. And though premises and presuppositions can direct trains of inference in one direction or another, they are also often hidden and not immediately identifiable. In any matter of controversy or in any complex matter of public debate, there are likely to be hidden presuppositions. In such instances, it is necessary to identify those presuppositions and bring them out of hiding. If they are unsound, exposing them to scrutiny may eventually turn the tide of opinion on the issue at hand.

Here, it is worth noting that philosophy has a history of diminishing premises; how little can be assumed at the start of an argument. It is necessary to begin with a premise because logic and reason cannot operate in a vacuum. Yet the less assumed at the outset, the less danger there may be in finding out later that a false assumption had been made. Perhaps the most long-standing tendency in the history of Western philosophy in regard to premises has been to remove the supernatural. Initially, this may appear to be a prudent move. Why assume more than absolutely necessary?

Our first hint of real trouble is seen in how the four questions of origin, meaning, morality, and destiny are tackled. In principle, this set of questions should be answerable from any premise that purports to describe ultimate reality. The expectation brings to light a curious state of affairs: The modern approach is for science to deal with the origin question while the other three are assigned to religion and perhaps philosophy. What can be concluded from that? It is not as if the answer to the origin question has no bearing on answers to the others. Whatever answer is proposed for the origin question should inform the answers to the latter three. The latter three are surely as valid; ordinary, everyday experience demands that. But can answers to the latter be as completely detached from the answer to origin as religion is nowadays detached from science?

Philosophical subjects are admittedly rather murky, at least when compared to scientific subjects. There may also be a certain anxiety about discerning truth in the metaphysical realm: How can we be sure the answers will turn out the way we want them to? Yet the fact is that everyone lives out conscious or subconscious answers to the questions of origin, meaning, morality, and destiny. You can absorb answers from your upbringing or from the prevailing culture unfiltered, or you can build them on critical questioning and reasoned thought. It may be impractical to avoid operating on at least some unexamined assumptions, but these can be evaluated over time and either corrected or affirmed. If proper attention is given to how presuppositions stand by their own wits and how answers to the first of this set of four questions connect with experience related to the latter three, philosophy may not be as difficult as initial impressions may suggest.

How Do We Know Anything?

n philosophy, the study of knowledge is formally known as *epistemology*. This branch of philosophy asks, "How do we know what is true, and how can we be sure of it?" The question is still under investigation by philosophers. Nevertheless, from the previous chapter, we should be aware that the knowledge we have of anything beyond our immediate sensory perceptions has to stand on rational inference. Rational inference itself depends on there being a rational order behind whatever sensations are experienced so that from those sensations, conclusions may be drawn about the causes of those sensations. This opens the door to a huge realm of knowledge, knowledge of the way things are that is beyond our ability to experience directly. The rational ordering of the universe allows us to know about subatomic particles and the War of 1812.

How do we know that the planets revolve around the Sun and not the Earth? Though we may accept this, it cannot be known by direct observation. Remember how Copernicus arrived at this idea in thinking that the universe must be more mathematically rational and orderly than the complex system of cycles and epicycles that the earth-centered view requires. Today, we have further scientific observations (such as shifts in the positions of nearby stars against more remote stars) and knowledge of physical laws (the law of gravity) that confirm the heliocentric solar system. We may further accept that the observations and laws do confirm a heliocentric solar system without going through the inferential exercise ourselves. Yet, as counterintuitive as the idea has been historically, lines of inference between the scientific observations and laws and the heliocentric conclusion were, at one time, rigorously scrutinized. It was the belief that the universe is rationally ordered that permitted that conclusion to be drawn as to the real cause of the observed planetary movements.

Much of what we know about experience is not from our own direct experience. It comes to us from others. But in receiving someone else's testimony about what they experienced directly, it does come to us through the same sensory equipment, usually through our hearing or sight. And at some point the direct experience of what is reported to us came through the sensory equipment of others, the same sensory equipment as we have. We also have the capacity to memorize previous sensations and the inferences derived from them. Any knowledge of reality then is either going to work or already has worked off of sensory perception. And along with sensory perception is the idea—the assumption—that there is a rational order behind whatever it is out there that is producing the sensations experienced.

As far as the natural world is concerned, the assumption of its universal rationality has not been all that common in history and has only been seen arising in ancient Greece and in Renaissance Europe. The assumption was not sustained in ancient Greece; in Renaissance Europe, it was sustained by a certain religious culture that was comfortable with it. The point is that the universal rationality of nature is an assumption, and an assumption that does not need to be made and often has not. Because rationality is assumed, there are no independent paths to verify the means by which knowledge is acquired. All one can do is assume that the world is rational and then proceed to draw inferences. The reliability of whatever knowledge we have about the world turns out to be more a matter of acceptance than proof. You really cannot *prove* knowledge of anything or that the sensory means of acquiring knowledge are at all reliable.[3] Thus faith is at least as important an aspect of existence as reason. There are certain prerequisites or presumptions necessary for obtaining any knowledge of reality at all. These are three:

1. An objective reality exists outside of perception—that is, independent of the mind.
2. A lawful regularity exists in this reality—repeatable, predictable cause and effect.
3. This lawful regularity is inherently rational and may be rationally perceived.[4]

These must be accepted without proof. Not one of these presumptions proceeds from either logical or empirical grounds. Collectively, they amount to faith in the rationality of all material reality. They are also essential operating presumptions of science. Science can get nowhere without them. Some scientists characterize faith as groundless belief, not stopping to consider what science is grounded upon. Some of the brightest scientific minds try desperately to hurl faith out of the world, both their world and everyone else's, but it cannot be done without removing the ground from under science at the same time.

Awareness of these presumptions is unlikely without some philosophical education, and so seldom is anyone conscious that there is faith involved in science. Thus, it is easy to hurl faith out or to at least put a stop to any more of it, and the means by which that is done is called philosophical naturalism.

Philosophical Naturalism

Science applies naturalism as a methodology but not necessarily as a philosophy. As a methodology, it is simply this: Let us see how far we can get with only natural causes to explain the operations of nature. Appeal to supernatural causes, though they may exist, is excluded by the methodology. Whether nonscientists agree with that or not, it is the practice of modern science. Naturalism as a philosophy is rooted in the scientific approach to knowledge, but it takes one further step. Naturalism as a philosophy is this: Natural causes are the only causes that exist. This means that nature is self-existent and that there is nothing in all reality that may be considered not a part of nature or independent of nature. A supernatural realm by definition does not exist; the possibility of a creator is eliminated. It also implies that causes observed in nature are random and without any sense of purpose or direction. Naturalism is well expressed in George Gaylord Simpson's often quoted line: "Man is the result of a purposeless and natural process that did not have him in mind."[5]

In practice, the ability to discern the line between naturalism as a methodology and naturalism as a philosophy is unlikely unless one is already well informed about the difference. Scientific organizations and scientists who argue evolution nearly always adopt and imply the philosophy in addition to the methodology. Asking whether philosophical naturalism is assumed in the context of any scientific pronouncement is therefore a necessity. (Philosophical naturalism should not be confused with natural philosophy, the prescience study of nature discussed in chapter 1.) We saw in previous chapters how naturalism under the pretense of science snatched jurisdiction over origins from the supernatural and from theology. Though consistently portrayed as scientific progress in the field of natural history and origins, this seizure was rather a philosophical and ideological substitution. And because evolutionary science is practically indistinguishable from philosophical naturalism, it thereby becomes the logical ground for atheism. Another way to state this is that evolutionary science is the "golden calf" of atheism.

The legitimate authority of the scientific method has thus been cleverly wielded to discredit the supernatural. A stipulation in the practice of science is put forward as if it were a conclusion. Many assume that the mystery of the origin of the universe, Earth, and all life-forms has been solved by evolutionary science. Nothing could be further from the truth, but this is what is taught in the public schools and by museums of science and natural history. Yet whenever this claim over origin is intelligently challenged, one finds that philosophical naturalism is suddenly replaced with the methodological position, that science does not concern itself with ultimate origin or what may or may not be behind the laws of nature. When the challenge disappears, the philosophical position is quietly reassumed. The distinction is very subtle, and the substitution is where much confusion arises and why it is so difficult to make any sense out of the evolution-creation controversy.

PHILOSOPHICAL RELATIVISM

Philosophical naturalism and atheism lead inevitably to preferential or relativistic views on questions of value, right and wrong, good and evil, and meaning. That is because if philosophical naturalism is true, mankind, as the highest intelligence, has no higher measure or standard of reference outside of himself for judging between right and wrong, good and evil, and meaningful and meaningless. If there is no fixed outside standard, then a question arises as to who or what should serve as the standard. There are plenty of sorry examples in history (with one exception) of someone stepping in to fill that role. An approach often taken in the academic world is to grant each individual the right to serve as his own standard. Decisions on matters involving prioritization of value, criteria for determining right and wrong, and what meaning or purpose should be pursued become subjective, matters of individual preference. There can be no acknowledgment that objective meaning and value or that fixed coordinates of right and wrong exist. That is philosophical relativism.

I happened to walk through a room once where a TV was playing some sort of invention program where a boy was trying to sell his latest invention to a skeptical audience. He declared, "Hey, if it wasn't a good idea, I wouldn't have thought of it!" Now what he literally was saying was that it would not be possible for him to think up anything that was not good or that goodness was somehow an intrinsic property of his. He removed the possibility of a conflict between his thinking and goodness. Such an assertion is an assertion associated with deity, and deities effectively are what philosophical relativism produces.

Relativism is appealing; the initial thrill of freedom from ethical or moral injunctions and the desire for autonomous self-determination is powerful. Relativism is at home with atheism because freedom from responsibility and accountability for one's speech and actions is total if there are no supernatural reference points for value, meaning, or goodness. The academic freedom to promote philosophies such as relativism seems unlimited. Yet note that academic freedom would itself seem to be an objective value, one to be "advocated in all circumstances." If this freedom were to be denied, the relativists would need to temporarily suspend their philosophy in order to contend against the censorship of it. To argue on behalf of a value such as academic freedom assumes that others might be persuaded by an appeal and any appeal has in mind something objective.

Other "isms" also proceed from naturalism, such as determinism (the absence of free will) and subjectivism (the absence of objective knowledge). In contrast to these, however, relativism actually produces consequences. It is not so much that anyone lives by relativism consistently. Chiefly, the application of relativism is pragmatic: As long as no one is harmed, why should certain practices or lifestyles be stigmatized? Any thought that harmful consequences might not immediately be apparent is disregarded. Yet in a complex world, consequences of relativistic trends are often not at all foreseeable. Strategies employed in attempt to remove cultural impediments to certain traditionally questionable practices and lifestyles involve the elevation of one value in order to suppress another. A reprioritization of values is precisely the intent of the secular humanism described in chapter 4 of this book. Examples now encoded in law include how the value of life was a few decades ago reranked below the value of privacy in regard to the unborn and how the value of the individual not too long before that began to be reranked above the value of marriage by making divorce easier. With a reprioritization of values, the landscape of right and wrong necessarily changes.[6]

AN UNLIKELY TARGET OF SKEPTICISM

Objectivity claims of science have been a subject of debate since the scientific revolution. To what extent can science claim that what it knows about nature is true about nature—that is, true with a capital "T"—and not merely the subjective interpretation of observations and experimental results? In particular, since at least the middle of the twentieth century, science has been the target of doubt from an intellectual trend that more recently has come to be identified with the term *postmodernism*. This challenge to science has arisen in humanities departments of universities and involves a perception that even scientific knowledge is not the true knowledge it claims to be. Postmodernism encompasses trends in art, architecture, and literature. In literature, the idea is that words convey no objective meaning but instead have subjective effects on people. If words do not have objective meaning, then scientific claims do not either.

That words have no meaning is not exactly how the postmodern critique of science is fashioned (though it might just as well be). The critique is rather presented as the inability of science to separate itself from the subjectivity of the scientist. The scientist as well as anyone else sees even the natural world through personal prejudices as well as filters and biases that are products of his or her cultural environment, same as with anyone else. The postmodern claim is that the assumptions that go into the sorting and interpretation of evidence are arrived at subjectively. To some extent, this is true. Subjectivity is what we have seen in some of the history of science covered by chapter 5 of this book and why scientific theories have sometimes been overthrown. The fallback of scientists to the postmodern threat is, of course, that science works, and works astonishingly well.

I see the postmodern denial of scientific knowledge as an extreme consequence of the philosophical naturalism inspired by evolutionary science returning in a round-about fashion to haunt science. The denial of scientific knowledge is justified because, as we began to see in the previous chapter, naturalism supplies no backing for objectivity. In effect, postmodernism has called the bluff of evolutionary science in that evolutionary science supplies no account for the origin or validity of rational inference and objective thought. The postmodern trend is the type of trend that must eventually arise when a transcendent basis of rational inference and objective thought is abandoned, and it must be abandoned if nothing exists outside of nature.

In similar fashion, postmodernism has concluded—reason and science having failed to supply answers to the meaning of existence—that there truly is no meaning to existence. Wherever the transcendent order that upholds rational inference and objective thought is swept away, any fixed coordinates for meaning and value that might have stood beside it are likewise swept away. But the postmodernist onslaught against truth and meaning must be done with stealth. This is necessary so that no one would think to apply postmodernist critical methods to postmodernists' own discourse.

THE SELF-DEFEATING PREMISE

Thus, the essence of postmodernism, that truth and meaning are empty ideas, is awfully problematic if postmodernists wish to be taken seriously—that is, in a true and meaningful fashion. Whenever a claim is advanced about what is real and true, it needs to satisfy its own test, its own criterion. Descartes began with the premise that only what is rationally defendable is real. But the principle of excluding all but what is rationally defendable is not in itself rationally defendable. This is because the validity of rational argument must be assumed upfront in order to begin to rationally defend the validity of rational argument.[7] Descartes had to begin somewhere, and he began with reason. But reason cannot defend the validity of reason should reason itself fall under scrutiny.

Philosopher David Hume (1711–76) began with the idea that all knowledge arrives in the mind through sensory perception. But then the knowledge that all knowledge is sensory must also have arrived in the mind through sensory perception. (It did not.) His claim about knowledge breaks down by its own criterion. Philosopher Immanuel Kant (1724–1804) on the subject of ultimate reality was agnostic, claiming that nothing of the immaterial can be known with certainty. But to say that is to claim to know with certainty something about matters involving immateriality. There can be no assurance that there is not yet some undiscovered path of insight into the immaterial realm without already having some fairly comprehensive knowledge about it.

From these examples, you may begin to see how an arbitrary premise can undercut its own ground. Philosopher Soren Kierkegaard (1813–55) tried to eliminate the need for any premise for metaphysical knowledge by altogether removing reason from it. He finished up the segregation of faith from reason that had for centuries been underway with his "leap of faith" philosophy wherein faith and beliefs about the supernatural no longer had any connection to truth and reason. This segregation is suspect in a similar way. The partitioning of faith and religious belief from truth and reason was put forward as truth, but was that truth also a leap of faith? If it was a leap of faith, how could he substantiate it or convince anyone else? Moreover, if the validity of reasoning must be assumed in the metaphysical, as it must be in making sense of nature, all that was really accomplished was to post a "No Admittance" sign against reasoned inquiry. He thus left the metaphysical in total subjection to individual taste.

The premise of naturalism is perhaps not obviously self-defeating, though there are dangers. If matter and energy began from nothing, the problem arises that nothing to something is outside the realm of natural causation and is thereby nonnaturalistic and nonscientific. The big bang theory of the universe is constantly bumping up against this problem of nothing to something. To avoid the inconsistency, cosmologists propose that matter and energy have always existed, a proposition safely beyond the reach of being disproven scientifically or philosophically. Or with its denial of truth and meaning, postmodernism, as an offspring of naturalism, shows naturalism to be self-defeating by that association. Of course, philosophical naturalists flatly reject these implications of naturalism; they are incompatible with the idea of meaningful progress toward scientific truth. In order to shield themselves from the contradictions involved in having rejected propositional truth, postmodernists escape into personally felt experience as the measure of reality.

But if objective thought is included in the scope of inquiry, naturalism does become self-defeating. As much as naturalism is advanced as a proposition whose truthfulness is to be pondered, its standing as a proposition crumbles once it must be recognized as an inevitable effect of a natural process. And as much as naturalism, in its assertion about ultimate reality, extends as far beyond human experience as it is possible to reach (the audacity of a universal negative as regards the supernatural), its standing as a universal claim collapses once it must be seen in cause-

and-effect terms as an offshoot of sensory experience. The problem is that if naturalism is true, naturalism must itself be an effect of natural causes, but if an effect of natural causes, it cannot be objectively true. The capacity of the mind for objectivity cannot survive the reduction of the mind to a product of nature. Subjective thought is the only type of thought a system whose processes work inevitably and whose effects are neither affirmations nor denials can produce.

Argumentation Has Limits

We have seen how naturalism falls rather effortlessly out of the scientific frame of mind, how relativism is an inevitable result of naturalism, and how the relentless logic of premises can produce even such extremes as postmodernism. There should be little surprise as to how Western culture arrived at postmodern thought. Now the question might be this: Can it back out?

It has been noted that philosophical arguments on questions of ultimate reality appear awfully circular. What circular means is that you end up where you started; the presupposition is also the conclusion. It should be no surprise that presuppositions are chosen with a view to where they will lead. As far as the choice concerns the evolution-creation question, one leans either toward the natural or toward the supernatural. The limitation stems from the fact that you have to begin somewhere and there is no neutral ground. Once having leaned in one direction or the other, it does not seem likely that any argument of your own could send you over to the opposite position. If God is excluded from the presupposition, it seems unlikely that he could be restored to it at any point in subsequent argument. If one starts with God as one's premise, it seems unlikely that he could be dropped off as superfluous at any point.

Not only do arguments about ultimate reality appear circular, but when one is dealing with such matters, evidence seems to be merely a means of persuasion rather than the authority for decisive resolution on which position is true. When views on ultimate reality are at stake, each contender attempts to use whatever leverage can be gained from the pool of evidence, but what one is predisposed to accept is the *de facto* authority. This point has been confirmed repeatedly in every forum in the origin debate because presuppositions and premises involving matters of ultimate reality are never permanently overthrown by arguments from evidence. Debate continues still. It would seem that neither side can get a firm enough grip on ultimate reality from evidence to once and for all remove the other from further consideration.

The points on circularity and evidence are not an attempt to discredit reason or the utility of argument but rather to show why argument alone seldom convinces anyone in matters of philosophical debate. What one is predisposed to regard as true is the determining factor in one's premise. If there is any certainty here at all, it is that people do not like being boxed in by what they take to be someone else's logic. This is why prior commitment to a philosophical premise cannot generally be unseated by argument. What changes one's thought is a change in what one is predisposed to accept. Whenever a sound argument conveyed in a straightforward manner is rejected, this point is worth remembering. Conversion to a new point of view may begin as a sudden realization that another viewpoint possibly has merit. Or conversion may begin with a lingering doubt or uncertainty about a given position that is eventually resolved in favor of an alternate view. Limitations notwithstanding, reasoned argumentation is all there is for the rational mind to lock on to, so arguments must continue to be advanced for the cause of truth.

THE PHILOSOPHICAL DEAD-END

The theistic premise, the claim that there is a creator, is not self-defeating. Nor is it circular because something that categorically cannot be explained by any natural process is found—objective thought. It is possible to break the circle should one stumble upon the right kind of evidence. Beginning with objective thought or beginning anywhere but including objective thought within the circle of one's arguments, one arrives at the conclusion that something outside of nature exists. C. S. Lewis advanced this argument more than sixty years ago.[8] It is the one argument that naturalism cannot deny without removing its own grounds at the same time.

Besides Lewis's argument from thought, arguments such as that from first cause, the cosmological argument, and the argument from design in nature, the teleological argument, have been in use for centuries and continue to be used today. They are accepted by those who are not predisposed against a theistic premise. Yet both have failed to arrest the scientific onslaught against creation, and both run up against the ambivalence regarding the intent or purpose of any particular natural object or phenomenon. Naturalists are able to deny their validity without digging their own grave.

It remains to be seen whether or not the argument from objective thought might fare any better against naturalistic predispositions.

Notwithstanding the merits of these arguments and others that have historical precedent, philosophical arguments proving the existence of God run up against a wall: One can arrive at the conclusion that there is a god, but you cannot *know* God. It is also hard to coax day-to-day relevance out of philosophical arguments. Much less can trust and hope be generated out of "God" as a hollow conceptual shell. Philosophical arguments for the existence of God are useful as academic exercises, but they are dead-ends because they fall short in leading to satisfying answers to any of the four philosophical questions noted above.

A worse problem with dead-ends exists for philosophy that refuses a creator. If the intent of philosophy is to find the coherency that ties together the diversity experienced in the world, it has failed. For 2,400 years, the quest has continued with no apparent progress toward consensus on definitive answers to the topics of philosophical inquiry. It is certainly the case that many presuppositions have been explored. There has been success with posing questions and devising arguments. But with premises on uncertain ground, few out of the supply of answers produced has seen near the degree of acceptance that attends scientific theories. Much more than in science, philosopher after philosopher starts afresh, unsatisfied with the attainments of the previous generation.

The search for a premise that is not only not self-defeating but also does not lead to a dead-end goes on.

Chapter 10:
Philosophy 2

He established it; he did not create it a chaos,
he formed it to be inhabited ...
I did not speak in secret, in a land of darkness; I did not
say to the offspring of Jacob, "Seek me in chaos." I the
Lord, speak the truth; I declare what is right.
—Isaiah 45:18–19 RSV

hilosophy has left promises of treasures of wisdom and knowledge unfulfilled. The parade of attempts to arrive at satisfactory answers to questions of origin, meaning, morality, and destiny has not ceased. Great philosophers have propounded many intellectually appealing ideas through the centuries. Some have run their course, having exhausted their explanatory power or having stumbled on some point that was less obvious early on. Others will doubtless cycle their way back again in the future as it is recognized that many have been reappearances of ideas entertained in some form in the past. Still others, while remaining entrenched in culture, reveal their inadequacy as foundations on which to build certainty. From generation to generation, the search for coherency and meaning apart from Christ goes on. The apostle has a point in saying that these treasures are hidden.[1] Can we step away from the philosophical carnival without abandoning the rigor of logic and discipline of reason that philosophy brings to the table? Continuing exploration of what was touched on in the previous chapter, how philosophical presuppositions on the origin of humanity may be evaluated, is the subject of this chapter.

Scientific inquiry usually begins with observation, and following that precedent, it seems advisable to likewise begin with observation. But there is another category of empirical data that science disregards, specifically evidence of intervention in the course of nature.

EVIDENCE OF INTERVENTION

The existence of intervention may be illustrated by the following scenario: Suppose a house were to appear where no house had before existed. Although houses are not naturally occurring objects, no one should find the appearance unusual; a properly equipped construction crew can cause that sort of thing to appear. The philosophical naturalist says that natural causes produced the construction crew—well and good. But how is it then that houses are not naturally occurring objects?

In posing this question, there is a definite resistance to move toward the category labeled supernatural. That term conjures up ideas of something beyond the human. Yet the category of natural does not seem entirely adequate. Hence, where does natural cause end and something beyond the boundaries of what we call natural causes begin? Consider Mount Rushmore National Monument, the heads of four US presidents chiseled out of the mount's granite. (We shall overlook the monument's new visitor center.) Which of the following three syllogisms is true?

Unguided natural causes do not and cannot produce such a monument.
Man is exclusively an effect of unguided natural causes.
Therefore, man cannot produce such monuments.

Unguided natural causes do not and cannot produce such a monument.
Man is *not* exclusively an effect of unguided natural causes.
Therefore, it may be possible for man to produce such monuments.

Unguided natural causes do not and cannot produce such a monument.
Man is exclusively an effect of unguided natural causes.

Therefore, man cannot produce such monuments, though perhaps with assistance outside of unguided natural causes he could.

This exercise illustrates a contradiction that the philosophical naturalist must suppress. Any natural object that is found in a state that is not its natural state is evidence of manipulation by something beyond nature. A carved flint is not the flint's natural state; a washed dog is not the dog's natural state; a trimmed hedge is not the hedge's natural state. The tacit exclusion given under a scientific heading to human intervention in the world should never be granted universally. Interventions in the operations of nature are the empirical evidence that philosophical naturalism cannot confront without exposing an internal contradiction. Something outside of nature exists.

It is clear that purposeful disruptions, alterations, and reconfigurations are imposed on the natural world. These disruptions and alterations of the otherwise normal operations of nature are effects that would never be produced by nature left to itself.[2] It is understood that the mind is the channel through which meaning and purpose enter the natural world. Intentions of the mind originate outside the realm of nature; nature does not contain intentions. The mind is free to impose its intentions because within certain limits and under favorable conditions, it is not bound by the natural world. More than natural causes are involved; an exercise of free will with intent is involved. There is no need to extend this argument beyond the human mind to say (in not arising from natural causes) what this free will with intent does arise from. All that is necessary is the recognition that absent an external power of sorts, no disruptions, alterations, or reconfigurations (it would not be a stretch to call them designs) could be seen in nature. However, in nearly all cases, everyone is instantly able to distinguish man-made objects from naturally occurring ones. Intelligent design is immediately apparent.

In philosophy, freedom of will is regarded as a problem. Pick up any introductory book on philosophy, and you will find it addressed as though it were a problem: Is free will real or an illusion? Are human beings deterministic in that every intention has a cause-and-effect explanation in theory, or contrary to the premise of naturalism, is there something about their behavior that in principle can never be scientifically predicted? It is philosophically impossible to prove that people in general have free will or that they lack it. But individuals should be able to determine if they themselves have it. I might not argue with someone who claimed to have no free will. But someone asserting that I myself did not have it would be another matter.

On the question of free will, it is interesting to contemplate the holding of responsibility. If there really is no free will at all, "Who is responsible for this?" is not a question that one would think should pop up as often as it does. The quest for a scientific answer would instead be the common response to an unsightly state of affairs. But a scientific answer would miss the point of the question. It thus appears that individuals implicitly assume that they have free will if only because they demand personal answers and not scientific ones from others. For this reason, I do not believe that the absence of free will is a viable philosophical position.[3]

Because nature operates according to set laws and is predictable to a large degree, there is a certainty regarding the eventual outcome of any intended disruption, alteration, or reconfiguration. Some of these redirections in the course of nature are temporary (e.g., the washed dog or the trimmed hedge). Others are more permanent (e.g., the house and Mt. Rushmore National Monument). All have potential significance. Redirections may have greater significance if they cannot easily be undone or undone at all (such as saving or taking someone's life). If redirections in the course of nature have significance, can some redirections be thought better or worse than others? If so, better or worse with respect to what? (Many things that are said and done are regrettable or commendable depending on how they are viewed or who views them.) More generally, are there means of weighing different redirections of nature and deciding between them? It is the type of question often raised.

THE REALITY OF GOODNESS

Science inquires about matters of fact, not of value. This statement is affirmed by scientists and is one with which no disagreement is expressed. In a philosophical system where scientific methods are viewed as the only means of obtaining knowledge, value judgments cannot therefore be regarded as knowledge. Nonetheless, the very fact that advocates of scientific origin theories so vigorously defend them suspiciously hints of the value they attach to them. Can they justify that value according to their philosophy? And where does the philosophy of naturalism leave matters of value and questions of right and wrong and good and evil?[4]

Because naturalism draws its justification from the view that scientific methods are the only means of obtaining knowledge, value judgments likewise cannot be regarded as knowledge in naturalism. Judgments about right and wrong are based not on any objective sense of value and goodness but are instead motivated by attitudes and feelings. Judgments of this sort say nothing regarding objective fact. If there are any facts involved at all, they are facts only about the person expressing a judgment and not facts about the value or goodness of the object or action being commented on. People have preferences regarding colors and flavors. Attitudes and feelings about right and wrong likewise reflect preferences. People normally do not argue over whether green is inherently better than blue or chocolate of more value than vanilla. Likewise is the logical implication of naturalism: It should be pointless to argue over matters of right and wrong or good and evil. Yet it is highly doubtful that parties involved in a dispute over values and what is the right course of action in a situation would ever confirm that all they are disagreeing about is an expression of either one or the other's attitude or feeling rather than a matter of objective fact.

A common escape from the namby-pamby philosophy of subjective (or relative) values is to adopt the position that matters of right and wrong are the result of cultural indoctrination, of conforming pressures by a community and its traditions. A localized pattern of behavior is said to emerge in order to achieve a measure of collective harmony or common good. But then how does a prospective community arrive at a position on right and wrong in the first place where one person's idea of evil may well be someone else's idea of good? At some point, someone, or perhaps a majority, would need to step in and impose some rules over everyone else. Otherwise, there might be an endless competition and confusion over what is right and what is wrong. But that is not the point on which the matter turns. We cannot grant, even for an instant, that a community and its traditions is the source of standards of right and wrong without encountering something higher. That "something higher" is, in this case, collective harmony, which has been smuggled in to become the objective standard. Collective harmony becomes the objective measure against which ideas about good and right are cast. Value is associated with the organizing of a community—a concept formed from two words: *common* and *unity*. But why community? Why not rather individual autonomy? Why should a state of continual competition of ideas of right and good not emerge? Is not competition what produced the edge for survival? Like it or not, an objective value has been introduced, namely collective harmony, one against which all conduct would be weighed.

The need of the naturalist is to suppress the contradiction between, on the one hand, the logic of the naturalistic premise that drives one to conclude that there is no objectivity as regards values and, on the other, the fact that people take their judgments of value and goodness seriously. That is a tall order. What the above paragraph shows is that one escape, the priority of culture and community over individual autonomy, is really no escape at all. An overarching value is buried in it. On the opposite view, assent to the existence of a supernatural ordering for values is not boxed in by the culture versus the individual dilemma. Reality is more complicated than what the cultural indoctrination or conforming pressures of community scenarios imply; collective harmony can coexist with individual autonomy to some extent, and both are worthy values. The supernaturalist need not agree to a supremacy of culture and community and will readily, even happily appeal to still higher and more objective values in case of conflict between community and individual. In some instances, it is the right of the individual that would be sustained. The naturalist cannot appeal to higher ideals without imperiling the view that value judgments are nothing more than personal or social preferences. All that remains are coercion- and domination-styled rules of community and there should be no pretenses about harmony or common good.

An evolutionary account may seem to be the answer, that human minds have been equipped with a sense of values, hardwired as it were, as an outcome of natural selection. And yes, it is true. Evolutionary theory may well supply an account for the arrival of value judgments as instinctive responses to situations. A behavior modifying overlay in certain of the functions of the brain, say, was found to be advantageous for the survival of a population. Or else a capacity for learned or conditioned response might be found advantageous for survival such that individuals in a population can be reconditioned in their responses as different situations arise in the environment. But "advantageous for survival" is predicated on there being value in survival. Is the theory itself not being overtaken by a concern over survival? Why should survival be advantageous? Or why should anything try to survive? And if evolutionary theory includes escalation from lower to higher forms, it is not merely survival but the value of progress that has overtaken it. But as design must be denied in evolutionary theories of origin, value must be denied as well. And if a value-laden theory of origins is denied, the question of how ideas about value arise, ideas that everyone accepts, remains as unanswered as the question of how design arises. For the naturalist, it is either that objective value is implicit in the theory or objective value is erroneous; neither position is sustainable.

Besides the presence of hidden values in certain theories about the origin of values predicated on naturalism, there is the fact that nothing is so remote from instinctive or conditioned response as rational inference. It matters not whether responses are biologically produced as instincts by natural selection or are socially conditioned by community pressures to conform or learned by cultural indoctrination. Responses geared toward survival (or else social acceptance) and rational insights into things beyond themselves are two different mental activities. It may be noted that value judgment stands in a position similar to that of rational inference: The discerning of value is as remote from instinctive or conditioned response as anything can be. Value judgments are likewise insights of a sort into things beyond themselves. They are not ultimately resolved or settled by indoctrination or conforming pressure. They may be intuitive perhaps but not instinctive. The rightness or wrongness of a value judgment is akin to the trueness or falseness of an inference. In either case, an instinctive, conditioned, or learned response throws no light at all on the matter at hand. The rightness of a valuation or the trueness of an inference must be established by some means outside of nature because nature has no yardstick of rightness or trueness in it.

The Bible informs us that to pursue knowledge of good and evil on our own terms is to be alienated from reality, in fact, from God himself. That is the message of Genesis 3. Reason alone cannot arrive at knowledge of good and evil because reason cannot supply its own premise. We are confronted with the necessity for proposing a criterion for defining goodness. But what is the criterion for choosing a criterion? Reason can only work off of something presumed at the start. If the history of ideas tells us anything at all, it tells us that without a fixed premise, reason can justify anything. What atrocities have the exalted nineteenth-century ideologies of communism and fascism *rationally* justified in the twentieth century? Many people say that religion also has been the cause of many atrocities. Nevertheless, one needs to ask whether such atrocities are the logical outworking of a religion's principles or the result of religion commandeered by an agenda that has nothing to do with those principles. If the latter, atrocities attributable to religion are then merely further instances of the sort of thing that results when good and evil are defined on our own terms. If the former, where atrocities are a logical outworking, then that religion is rightly judged.[5]

In contrast to the subjectivization of value and goodness, a transcendent origin stations value and goodness as objective and invariant. Judgments about good and evil stem from senses of value that are not merely matters of majority opinion or individual whim. In Western culture, moral reasoning, if not undergirded by, is at least attended by the inestimable value of every person, a deduction from their bearing the likeness of the Creator.[6] The Ten Commandments, after establishing the reverence due the Creator, go on to affirm that each person's life is inviolable (Thou shall not murder), each person's trust is inviolable (Thou shall not commit adultery), their property is inviolable (Thou shall not steal or covet), and each person's name is inviolable (Thou shall not bear false witness).[7] And in reference to the Two Great Commandments,[8] everyone is worthy of their neighbor's concern for their welfare because they, as with their neighbor, bear the likeness of the Creator, who is to be loved. As much as these biblical premises for moral reasoning are now stricken from public view, no one denies their goodness.

THE ARGUMENT FROM EVIL

"There is too much evil in the world, so how can a good God exist?" It may come as a surprise that this challenge, the argument from evil, adds more certainty to the existence of God than it takes away. This is because when you say that evil exists and that God should be good, you have already assumed that good and evil do, in fact, exist and that there is a means to distinguish one from the other. You take for granted that when you say "This is good," or, "That is evil," whoever you say that to uses the same measure to distinguish between them. One does not normally encounter skepticism over whether such a measure exists. Judgments are constantly being made about right and wrong things that someone else may be doing or may have done. The amazing thing is the expectation that such judgments will be accepted, and even more amazing is that they are!

There is more involved than individual preference, if only because implicit to the argument is the presumption that the existence of evil in the world *is* a real problem. There is no escaping the recognition that using the presence of evil to argue against the existence of God invokes criteria that transcend individuals. Even when the specific criterion to be applied becomes a matter of debate, the indignation and outrage often seen in debates does not suggest that these are mere subjective matters, such as a preference for chocolate over vanilla. While there is often confusion over which values or principles should take precedence in judging the goodness of a course of action, that confusion does not seem to diminish the number or intensity of commendations and denunciations about things someone else may be doing; nor does it diminish the idea that principles may be applied. The idea of principles and the principles themselves remain transcendent.[9]

Transcendent principles and transcendent criteria that distinguish between good and evil can only come from a transcendent source. But a transcendent source is that which they are attempting to disprove. The argument from evil is hazardous to itself. As C. S. Lewis observed:

> The defiance of the good atheist hurled at an apparently ruthless and idiotic cosmos is really an unconscious homage to something in or behind that cosmos which he recognizes as infinitely valuable and authoritative: for if mercy and justice were really only private whims of his own with no objective or impersonal roots, and if he realized this, he could not go on being indignant. The fact that he arraigns heaven itself for disregarding them means that at some level of his mind he knows they are enthroned in a higher heaven still.[10]

Implicit to this challenge to the existence of God is a demand for an explanation of the evil observed. Does this demand make any sense in a context of philosophical naturalism? The one thing that cannot exist if philosophical naturalism is true is a transcendent and objective standard of good and evil. The challenge always breaks down on this point: Take God away, and good and evil vanish as well. We are not giving an explanation here for the existence of evil. That would require knowing the meaning of existence and is a topic beyond the scope of this book.[11] What we are saying is that if having established that God does not exist, the only remaining possibility for an explanation of evil is whatever may be searched out from within the individual as to why certain happenings, motives, and actions are preferred over others.

Moreover, in talking about evil, we are not talking about the merely accidental or a result of ignorance. We are talking about evil saturated with malicious intent. There is no place for intent in a system of philosophical naturalism where all happenings are effects of natural causes. And the more emotionally charged this challenge to the existence of God is, the more outrage expressed by the skeptic over conditions in the world, the deeper the hole he digs for himself. Outrage can only be generated within an objective moral context. A bland, matter-of-fact demeanor is all that the skeptic can risk without undercutting his own ground. So much for the passionate force of the argument.

CONSISTENCY AND RELEVANCY

T he loss of scientific credibility in certain academic circles was previously discussed—how the acceptance that there is real scientific knowledge was displaced by the idea that nothing can be known with certainty. Postmodernism, as this trend was known, was internally inconsistent because to say that nothing can be known with certainty is itself something that is asserted to be known with certainty. Postmodernism was often aimed at the meaning of words. But to say that words have no certain meanings is a use of words under a presumption that they do have certain meanings. Anyone who would have you accept that presumption does not want you to think about it. Fortunately, this trend encounters continual resistance in scientific circles, as it is incompatible with meaningful progress in scientific knowledge. The postmodern threat to science seems to have retreated at present, but it illustrates well the general problem that arises for any ideology that would undermine objective truth and objective value: You cannot assert the objective truth and objective value of claims that undermine objectivity.

Yet that is just what the premise of naturalism does. Knowledge must be accounted for by natural causation. Any account of knowledge on other grounds is a violation of naturalism. But the stipulation that only natural causes exist leaves no room for the very thought process science depends on to infer knowledge from sensory perception. It is not possible to see (in the sense of *knowing*) how cerebral biochemistry enjoys special status above any other biochemistry as regards a capacity to produce knowledge. The mind cannot be in a causal relation to that which is known without being, at the same time, subject to that which is known (as an effect is subject to its cause). Objective knowledge cannot arise from nature because to know objectively is to know independent of any effects produced in the mind by that which is known. Objective knowledge is such that it exists whether or not the mind perceives it. It is, in a sense, out there already, not something that exists only in the mind. The real relation involved in knowing is a logical one, not causal, and the logical relation is extra-natural.

Even if it were possible to offer a causal account of knowledge, to do so renders an account on rational grounds superfluous. It is not necessary to consider reasons for an assertion or belief that is already shown to be the result of nonrational causes. In fact, it is the act of attributing an assertion or belief to causation, whether true or not, that becomes a prime justification for rejecting rational grounds. And once causation is established, however it may be established, the question as to how the truth of an assertion would ever be assured remains unanswerable. The logic of philosophical naturalism returns to haunt it; the assertion that naturalism is objective knowledge hollows out its own ground. Lewis summarily describes the root contradiction of naturalism this way:

> It follows that no account of the universe can be true unless that account leaves it possible for our thinking to be a real insight. A theory which explained everything else in the whole universe but which made it impossible to believe that our thinking was valid, would be utterly out of court. For that theory would itself have been reached by thinking, and if thinking is not valid that theory would, of course, be itself demolished.[12]

While naturalism may not be as easy a dismissal as postmodernism, the difference between them is not one of kind. The problem of perception is what helps to keep them both afloat. It is easy to overlook rational inference, the tool of scientific inquiry, when it is in use. You do not focus your awareness on the hammer but on the nail. The tool does not become an object of inquiry unless you take a step back to include it among those things that need to be explained. If the tool of postmodernism is language, the tool of science is rational inference. But if science is to become the exclusive producer of knowledge as certain advocates seem to want to have it be, our thinking must be included among those things that need to be explained. You cannot invalidate the tool with the theory that is supposed to explain it.[13]

Besides the test of internal consistency, the validity of a philosophical premise may be seen in its relevancy to common experience, whether it informs, validates and energizes common experience, or collapses and nullifies it. The questions of meaning, morality, and destiny were mentioned earlier: Why am I here? How should I live? Where am I going? An evolutionary origin denies the relevance of these questions because it supplies no answers. It does not offer guidance on finding purpose or meaning. It supplies no framework for deciding on good and right courses of action. It does not offer an account for the persistency of hope that physical death is somehow not the end. Science acknowledges its lack of answers implicitly when it cedes the subjects to religion. Yet in its preemptory dominion over knowledge, science retains under its jurisdiction the one answer that would likely inform answers to these questions, the answer to "Where did I come from?"

While everyone acts as if there are answers to the questions of meaning, morality, and destiny, it may be reasonable to ask if objective answers do exist. There are, however, further indications from common, everyday experience that the answers are not merely subjective preferences. Purposes are judged according to their honorableness. Determinations of good and right are directed by conscience. Hope persists across cultures and generations that physical death is not the end despite an apparent lack of means to empirically verify any grounds for such hope. Consideration of answers to these questions is beyond the scope of this book. But if ever objective answers to meaning, morality, and destiny would be found, they would be found where the Creator would be found, obviously not from the study of nature.

Perhaps the most telling test of relevancy goes back to the assumptions required for knowledge of any sort—that an objective reality exists outside of perception, that a lawful regularity exists in that reality, and that that lawful regularity is inherently rational. How much trust is placed in these assumptions by scientists? By everyone? How can anyone be justified in placing their trust in unprovable assumptions? The point is that the creation premise, in not casting away the transcendent, has a potential to validate that trust. On a premise of naturalism, there is nothing more to say—not exactly the most informative complement to our unquenchable sense of a rational world. Reasoning both to and from the creation premise is straightforward in its consistency and relevancy; the Creator is necessarily rational, purposeful, and good, though yes, there may be more to know about said Creator.

Concluding Remarks

Science over religion, evolution over creation, reason over faith. Great minds of science and philosophy have tried hard. Yet science, evolution, and reason fall short in supplying a satisfactory premise on which to build a coherent philosophy. It is fine for scientists to stipulate that only natural causes are permitted in the practice of science; they have the right to define the terms of practice of their profession. But when a stipulation is put forward as a conclusion, there is a serious problem. With that move, science displaces other sources of knowledge, positioning itself as the only source of knowledge.

Granted, science stands apart in a distinctive way. While a multitude of religions contend for their particular version of a something beyond nature, science stands alone as objective knowledge of the natural world. There is an aura of neutrality, but there is no neutrality because either a supernatural realm exists or it does not. What has been covered to this point says that there is. A rigorous application of philosophical methods says that a supernatural realm exists because rationality, purpose, and goodness exist. Any presupposition or theory regarding origins must at least exhibit logical consistency with and relevance to these three aspects of experience. There remains yet to introduce empirical credentials of the supernatural. This we will do in the last two chapters.

Finally, there is a game of shifting shadows. While conveying an aura of neutrality, the evolutionary premise-conclusion is assumed without question. The cult aspect of culture is at work here; no one dares to question the power of the scientific method that has reaped the wonders of learning and technology. Whenever anyone gets close to unveiling the faulty premise of naturalism, evolutionary scientists suddenly shift to an agnostic stance that makes no assertion about an ultimate premise. This is the methodological position, that naturalism is merely a method, not a philosophy. Once the threat has dissipated, philosophical naturalism is reassumed. The game is played on an unsuspecting public who largely believes in the beyond but has not been taught the intellectual rigor needed to probe the murky world of philosophical presuppositions. And as long as public education is ruled by those who subscribe to the ideology of naturalism and its associated religion of secular humanism, this state of ignorance will persist.

The world is not a place of darkness and chaos. It is possible to know what is true and right. The causes that science limits itself to cannot account for rationality, purpose, and goodness, so we are constrained to look toward the supernatural to explain them. That conclusion may be about as much as philosophy by itself can achieve.

Chapter 11:
Revelation

I am the first and I am the last; apart from me there is no god.
—Isaiah 44:6 NIV

The words of the first and the last, who died and came to life.
—Revelation 2:8 RSV

Aghast at science's conquests in the second half of the nineteenth century, a large contingent of the clergy began to deemphasize divine revelation, even to the point of foregoing it as the source of their knowledge and authority. They then attempted to get by on what was not objectionable to science. Some even reformulated their religion exclusively on humanitarian ideals; they preached economic and social equality rather than individual liberty and spiritual conversion. As the scientific advance continued, theology in general was in a particularly vulnerable position because people thought the existence of God was proven by nature. Nature proved the existence of God, and God was then assumed to speak through the Bible. Science obviously challenged the thinking that nature proved the existence of God. The idea that the Bible stood on its own merits did not appear to be a matter of general awareness. Otherwise, the theological deference to science might not have occurred. Nineteenth-century higher criticism sealed the trend with an assumption that the ancient writings were not divinely inspired and that the reports of miracles were false. That assumption was presented as if it were (after what should by now be a familiar pattern) a conclusion.

It seems absurd to the scientific mind to attribute our existence to miracles and supernatural entities that cannot be assessed by empirical means. On the other hand, to think that the universe, earth, and all living things came about as the result of happenstance, by a mayhem of natural causes, seems equally absurd. How we arrived here and why are arguably the most important questions anyone can ask. Without fixed answers, purpose is cut loose; without purpose, values are formless and void. Not everyone always has a clear sense of purpose, but at least no one consistently lives valuelessly. There does seem to be signs of fixed answers. Even evolutionary theory, contrary to the interest of philosophical consistency, is deemed valuable as shown by science's defense of it. At this point, we know that something must have preceded nature, and neither science nor philosophy has found it.

While that which preceded nature may be high on the list of academic questions, the more troubling is what follows it. How desperately do the living hang on to life and hope! The bigger mystery is what happens after you die. Is that the end, or is there more? Is hope real or empty? If religion is of any value at all, it must supply a verifiable answer. Nature, reason, and conscience are unable to answer this. Nature is indifferent. Without fixed premises, reason is apt to be swept along by anything. Conscience only declares one's unworthiness. What is left?

Yet religion is divisive. Beliefs seem beyond rational discourse. As much as they are avoided, the problems involved in grounding premises of a universal character in philosophy do not really go away in religion.[1] Sometimes a suggestion is made that the various religions are just different paths to a common understanding. But there does not seem to be any indication that anything more than the most superficial common understanding has ever been attained. You do not need to know very much about religions to identify them with their rituals, traditions, beliefs, and practices. There is no mistaking one for another. What could possibly cause adherents of one religion to merge their practices, traditions, and teachings with those of another? How could Judaism and Christianity ever even agree on what day of the week to regard above others? The divisiveness of religion and the perception that there are no rational means to assess the validity of these belief systems would seem to justify dismissing them.

There is, however, an instance of agreement between religions or at least between their writings in that one book contains the scripts of two of them. In view of the divisive nature of religious belief and practice, it is extraordinary that foundational writings of two major world religions should end up in the same book. There is more involved here than religious belief and practice; the respective beliefs and practices of the two belief systems are not compatible enough to stick their texts together. What is seen is that neither collection of writings by itself is a stand-alone document. In the first, while a religious tradition is established, a future is anticipated that is not contained within itself, and it remains forward-looking to its end. The second finds a template of sorts for itself embedded in the first in that the events it records satisfied the first's long-established anticipations. The specifics of the second text could not have been predicted from the first text by itself. But through study of both texts, the linkage is seen to be comprehensive. The unfinished business of the first is wrapped up by the second. Each of the two collections of writings is made complete by the other, a fact in view of the nature of religious belief that cannot be dismissed as a human invention. The pairing is unique; the underlying compatibility cannot be denied. A moment's consideration leads one to realize that the Bible's seal of authenticity in its claim to divine authorship begins with a transcendence of religion.[2]

Skeptics always have the miracles card ready to play. Yes, reports of miracles are rightly regarded with a high degree of suspicion. For the sake of credibility, if a proposition regarding the secrets of life and death is to be put forward, it would be better to put forward what is at least somewhat sustainable within the realm of natural causes. Even a proposition that is impossible to either prove or disprove, such as a universal spiritual presence, reincarnation, or the idea that life is merely an illusion, may be more believable than a miracle. Why should disbelief be prompted by proposing the incredible? Nevertheless, a proposition was advanced that was highly vulnerable to being disproved at the time. The New Testament's writers recorded that the Old Testament's Messiah was raised from the dead in accordance with the Old Testament's predictions.[3] As a human design, such a report reached beyond prudence. It would have been false because everyone knew that it was humanly impossible to reverse a transition from being alive to being dead. Yet one thing may be said: At least this incursion into the secrets of life and death had no arbitrary character but all the realism of a record of specific events in history.

But it was a record of specific events in history. The record of specific events was this: There was a man named Jesus. He was not a myth or a legend. The leaders of the Jews conspired to have him killed and when he died, he was dead; there was no uncertainty about that. The tomb in which he was buried was later found to be empty; there was no mix-up in tombs. And the reason it was found to be empty is because he had risen from the dead.

Notably, testimony of the resurrection was not opposed by correcting "erroneous" facts as to what had occurred or by countering "deceitful" interpretations of the Old Testament prophecies about it. It was opposed by violence and slander. This type of opposition indicates that it was not motivated by a desire to set the record straight but rather to silence the testimony. The opposition was organized at once by the very Jews whose scrolls foretold the miracle. The fact that the fulfillment of the Old Testament law and prophets was rejected by the Jewish authorities and scribes means that the purpose that unites the Old and New testaments was never their planning to begin with. What then became the focal point of history is evidence of a power greater than that which opposed it: "The stone which the builders rejected" had "become the head of the corner," and it was the Lord's doing.[4]

The report of the resurrection soon spread throughout the known world. It was not a fabricated story; there were too many eyewitnesses of the events surrounding the resurrection still living. The eyewitnesses were unanimous. No contemporary dissenting accounts exist. The record and explanation of events were written down within the lifetime of those witnesses; and as many copies were distributed, it was thereafter impossible to enrich their content. The charge that it was all a myth or a legend cannot be sustained in the face of consistent eyewitness testimony and contemporaneous dispersal. The resurrection of Jesus Christ from the dead remains one of the most verifiable facts of history. Every attempt to explain away the straightforward record of the New Testament on this claim has been a failing exercise resulting only in absurdities more elaborate than the supposed one to be done away with.[5]

The resurrection of Jesus Christ also poses a quandary with regard to the origin of life: If resurrected life is miraculous, as it doubtless must be, is life at its first somehow not? The difference between alive and dead can be perceived in an instant, yet what is behind that difference is shrouded in impenetrable mystery. To be alive is a miracle. And it should not be impossible to discern who has the first and last word in this matter.

Chapter 12:
Reason and Faith

*And Paul went in, as was his custom, and for three weeks he argued
[reasoned] with them from the scriptures, explaining and proving that
it was necessary for the Christ to suffer and to rise from the dead.*
—Acts 17:2–3 RSV

*No distrust made him waver concerning the promise of God,
but he grew strong in his faith as he gave glory to God, fully
convinced that God was able to do what he had promised.*
—Romans 4:20–21 RSV

Those who are wise in their own estimation are drawn to the
discourse of the foolish who say, "There is no god."[1] We note that
the statement, "There is no creator," is a particularly central theme in
college and university environments. Naturalism rules. Is there ever found
such paradoxes of wisdom and foolishness as the agnostic and atheistic
professors of philosophy and science?

Faith in a creator is portrayed as uncorrelated with reason and
hopelessly outdated in view of the triumphs of science and unworthy
of intellectual respectability. The association of empirical evidence and
objective reasoning with the scientific method has generated a perception
that science has exclusive claim to the entire realm of reason and rationality.
There now exists an implicit assumption in Western culture that knowledge
can only be attained by the methods of science—that is, by observation

and reason.[2] And because knowledge of the supernatural is not a subject of scientific inquiry, the inevitable conclusion is that no knowledge of anything supernatural is possible. Therefore, any beliefs regarding a creator and the supernatural must be accepted on "faith," faith being defined as unquestioned belief that requires neither proof nor evidence. The phrase "taken on faith" insinuates the acceptance of beliefs that cannot be verified as being true, and so truth has nothing at all to do with whether or not they are adopted.

If science has staked out its territory, faith has as its territory what remains—superstition and myth. We are thus assured that a belief maintained on faith is in an altogether different category than knowledge based on reason. This has become the popular sense of the term *faith*. The partitioning of faith and reason is the triumph of the rationalistic, naturalistic, and agnostic philosophy that began to arise during the Enlightenment era and has now come to rule Western culture.[3] The apparent occupation of the entire realm of reason and rationality by science has caused many believers in the Creator to think that they need to surrender to the scientific onslaught in order to retain any association with rationality at all. Science has been pleased to offer terms of surrender: If you wish to retain intellectual respectability, purge the supernatural content from your beliefs.

Does this castigation of faith have any validity? Is this portrayal of a separation between faith and reason a true picture of reality? Does the expulsion of faith from rationality make any sense? While enlightened academics contrast faith and reason, the scriptures contrast faith and sight.[4] Nothing is said or implied about any partitioning of faith from reason. The scriptures take the capacity for reason for granted. The strongest arguments advanced by the apostles in the New Testament to substantiate Jesus as the Messiah and his resurrection from the dead were not their consistent eyewitness testimony, as legally authoritative as such testimony is. Nor were they the confirming miracles that accompanied their gospel. Their strongest arguments were reasoned from Old Testament prophecy. Appeal to those writings was their customary practice in persuading audiences as documented in the book of Acts.[5] Time after time, the New Testament draws from the Old in support of its claims. The New Testament scriptures appeal to reason. They also appeal to faith: "And without faith it is impossible to please him. For whoever would draw near to God must believe that he exists and that he rewards those who seek him."[6] Reason and faith are at home with each other here.

The author of the New Testament book of Hebrews sets forth the scriptural definition of faith: "Now faith is the assurance of things hoped for, the conviction of things not seen."[7] Faith, according to the scriptures, is what sustains in one's mind the reality of what is hoped for and unseen. Faith is *not* how a belief is first arrived at; let this point sink in. Rather, faith is the ability to continue to believe in an unseen reality once established when the grounds for it *seem* remote or when sight would sweep it away. People want to think that faith becomes a substitute for reason where reason cannot be applied or when there is a failure to properly arrive at truth. That is not it. Faith is in the continuity, not the initiation of a belief. C. S. Lewis wrote, "The difficulty of such continuing to believe is constantly ignored or misunderstood in discussions of this subject."[8] How great is the tendency to settle back into the narrow boundaries of perception that defined what the New Testament calls the old nature. That is why the New Testament scriptures' appeal to faith is central to the nature of the new creature of spiritual awareness. None of the things associated with that spiritual inheritance can be seen by physical eyes; having been established as existing by reasoning from Old and New Testament scriptures, their reality must be sustained by faith.

The proper sense of faith needs to be set straight from the pressures of higher education that wish to make it seem crooked. The notion that some intellectual failure or breakdown has occurred wherever faith is found is false. That notion is a subtle attempt to malign true faith. The failure or breakdown, if there is one, would be on the part of reason to arrive at truth in the first place. If "the assurance of things hoped for and the conviction of things not seen" is a meaningful definition, the truth of what is hoped for and unseen must be verifiable. In the case of the scriptures, what is verifiable is the following:

1. The foundations of Christianity were written into the Old Testament long before Christianity ever appeared—conclusive evidence of the existence of one plan and one mind behind both testaments.

2. Christianity was established largely on the authority of the Old Testament in the face of Jewish opposition—conclusive evidence that the Old Testament was never a Jewish invention in the first place.

Yes, it is possible to apply faith to sustain belief in a claim that has not been verified as true; however, it does not have to be that way, and the scriptures do not require that faith be so applied.

Faith is never far away. We have the capacity for faith as well as reason. We have, for example, faith in ourselves. We undertake to do something that we have never done before. Because we believe that we can do it, we often can, and it gets done. It might be said that we "foresee" ourselves doing whatever is set in mind to be done. There is a confidence before the fact. There is no rational connection to knowledge, but there is a knowing still. It is clearly not reason, but that does not make it the antithesis of reason either.

We have the capacity for faith in someone else. The word for that is trust. If one in whom we have placed trust makes a promise, then we have faith that they will bring it to pass. Such faith may leave reason behind, though not far behind; it is well that a venture to trust be preceded by some evidence of trustworthiness. In the context of the scriptures, trust in God is no different than trust in another person. Faith in God is not portrayed as being in an altogether different category than faith in a fellow human being. At one time, he appeared as a fellow human being.[9] If, like anyone else, God proves himself faithful, we can trust him. The extension of the term *faith* to *faithful* or *faithfulness* well explains this sense of faith in distinction to the false sense of ungrounded belief. Where faithfulness is exhibited, trust can be generated.

If faith without reason is how the scientific mind considers religious beliefs, the accusation has validity in view of the multitudes of propositions believed without any attempt to verify whether they are true. A danger arises here because faith empowers the will to act. Faith produces, and faith achieves. How often has a faith picture, an image of a future reality in someone's mind, caused results to happen? The first circumnavigation of the globe and the first moon landing were results of faith. Individual accomplishments such as achieving education and career goals are generated by faith. These are examples of faith and reason working together. But what about faith without reason? If what is believed is not or cannot be reasoned out, what then? Faith without reason can produce extraordinarily irrational and immoral acts. For example, faith without reason can strap a bomb to itself, step onto a crowded bus, and blow itself up. The condemnations heaped on faith are indeed often justified when faith operates without reason.

What about reason without faith? Reason may discern a relevant truth, but unless there is something that generates action on it, leveraging that truth to affect future reality, it would seem to stop at mental assent. That is not all. Fears and desires often oppose the responses to situations that reason has determined are good and proper. In opposing temptation, reason has always been assumed, observes Lewis.[10] Yet reason, he explains in another place, is often found powerless against temptations. He goes on to draw this analogy: "When once passion takes part in the game, the human reason, unassisted by Grace [referring to divine grace, the response to which is faith], has about as much chance of retaining its hold on truths already gained as a snowflake has of retaining its consistency in the mouth of a blast furnace."[11] Reason may even be found justifying or excusing the responses that fears and desires produce. Reason is necessary, yet by itself, it produces nothing or, worse, becomes an excuser of ethical lapses. Something else is necessary, and that something else is faith. If reason gains for us a truth, faith sustains its reality against onslaughts of irrational doubts, fears, and desires and then finishes with action.

To complete the circle on the experience of faith, it is far more a response to integrity of character in its dependability than it ever is to propositional truth. Consider now these words from Ravi Zacharias on faith:

> One of the most startling things about life is that it does not start with reason and end with faith. It starts in childhood with faith and is sustained by either reasoning through that faith or by blindly leaving the reason for that faith unaddressed. The child's mind has a very limited capacity to inform it of the reason for its trust ... If over time that trust is tested, it will be the character of the parent that will either prove that trust wise or foolish. Faith is not bereft of reason.[12]

Is a childlike faith unjustified, unreasonable? Does it become unreasonable as the capacity for reason develops and can be exercised? To insist that reason always precede faith ignores the fact that between persons, the sequence often runs in reverse. Or, to reject faith at a point and replace it with reason causes relationships to wither. The either/or partitioning that stands faith in isolation to reason is unconstructive. The Father in heaven requires childlike faith,[13] and if you must, reason is not excluded. The Father's integrity of character is shown in the testimony of the prophets and apostles—in how the ancient promises were fulfilled in his sending the Savior, how the Savior made known the love of the Father, and how the Father grants the privilege of free decision on whether love will be reciprocated by trust. Wherever love is real between parent and child, it follows the same pattern.

How did the popular sense of faith arise? What lurks beneath the perpetual skepticism and misuse of reason to reject faith? Prior commitment to an evolutionary ideology may be involved here, but the direction in which the will is aimed is at the root. Faith will always be portrayed as credulity when the slightest ramification of belief in the supernatural does not conform to the vaguest preconception of how reality should be configured. The skeptic decries a lack of scientific evidence as justification for a refusal to confront a reality known by other means. He then triumphantly boasts that reason and science have conquered faith and superstition. But is there really a lack of evidence? The scriptures do not suggest that more evidence is always what leads to more faith. The disbelieving scribes and Pharisees were refused more on-the-spot evidence:

> Then certain of the scribes and of the Pharisees answered saying, "Master, we would see a sign from thee." But he answered and said unto them, "An evil and adulterous generation seeketh after a sign; and there shall no sign be given to it, but the sign of the prophet Jonah."[14]

And is not the demand for a sign a sign in itself? asks Zacharias.[15] The use of the mind by the skeptic to require more and more evidence for belief ignores the existence of the mind in the first place. If there is no reason and no mind at the root of existence, how can the escalation of demands for a reason make any sense?

The scientific requirements of the skeptic may be nothing more than a smoke screen to conceal a fear or contempt of the self-appraisal and self-sacrifice necessary in committing to truth. Where does one's will stand in relation to truth? Does it yield to truth, or is it subject to some other agenda? That is the question that needs to be answered before the truth of any proposition can be considered. As Zacharias has again insightfully ventured, "In any interplay or confrontation between a person and information, the first test is not the veracity of the information, but the truthfulness of the person."[16]

It appears that the world of higher learning has seriously corrupted the meaning of faith. Do not be taken in by this. While faith is to be contrasted with sight, it should never be divorced from reason. Reason and faith are both intrinsic attributes of personhood that should never be set in isolation from each other.

Summary
and Conclusion

The pronouncement of a separation between science and religion is the big gun that allows evolutionism to consistently prevail in confrontations with creation, whether in courts of law, public schools, or elsewhere. Hundreds of years of enlightened philosophic and scientific thought have partitioned off religion as unworthy of consideration as truth, the antithesis of science. It is the same partition as that stationed between reason and faith. Considerable fragmentation of what is seen to be religious authority has fed the perception. Science is thus guaranteed a high seat in public forums and public institutions while religion confronts an impenetrable barrier: Its entry into the public sphere is an unconstitutional violation of church-and-state separation. If it is the particulars of denominational religion, the barrier is a necessity. But the consequence is that evolutionism is welcomed as scientific truth while creation is held in disdain as unworthy of even being deemed false.

The singular fact on which the partition crumbles is that science depends on a rational world whose ordering is universal. The truth that science seeks to discover cannot exist without that ordering. But that ordering cannot be known by scientific inquiry; the reach of scientific observation and experimentation is limited. Scientists know that the rational ordering is consistent through all time and all space, but how do they know that? It can only be accepted and then maintained by faith. It is a religious proposition that, historically, only Christianity has affirmed. Absent continuing faith in a rationally ordered world, scientific inquiry would cease. The National Academy of Science's dividing line between scientific and religious thought dissolves on this observation; science and religion cannot be split. Attempt to enforce a division must in fairness assign scientific causes to scientific prerequisites. That attempt stumbles on the absurdity of casting faith in a universal rationality and the validity of scientific reasoning as somehow a product of natural causes. Indeed, the reverse is true: the validity of natural cause and effect proceeds from faith in a rational order and faith in scientific reasoning. But as science acknowledges only cause and effect, what is one supposed to conclude? The truth that scientists know is not a product of the causes science considers.

The evolutionary thought system is built on assumptions foreign to empirical methods: natural causes only in the unobserved past and naturally occurring specialization and ordered complexity. Uniformity of natural causes in the unobserved past functions well as an organizing principle in geology, biology, and astronomy, but its truth is not thereby known. Requirements necessary for the formation of fossils and the presence of C-14 in certain rocks have not been well integrated into uniformitarian geology. The conventional uniformitarian age of the universe does not appear to have been reconciled with the space-and-time relativity of Einstein. As much as the idea is loathed in science, the book of Genesis with its consequent age of creation and consequent worldwide flood catastrophist hypothesis on formation of fossil-bearing strata might remain a viable alternate assumption. The idea that biological systems increase in specialization and ordered complexity further unifies biological science, but that idea is not empirical either. Moreover, the evolutionary process was a process that is alleged to have been neither rational, purposeful, nor of value. It just happened. And if the attributes of rationality, purpose, and value are irrelevant to the process, then the process is irrelevant to the attributes. The question as to how such attributes arrived is unanswerable by this thought system.

A specialized method of inquiry, equipped to be applied only to a limited scope of problems, cannot be applied to everything. What evolutionary science has wrought is a serious plunge into redefining the nature of existence. There is a logical consequence in positioning every fact of existence on a scientific base. We do not know if there is meaning to life because science cannot discern meaning. While it is said, "We scientists have seen that everything can, in principle, be explained by natural cause and effect," no account is ever offered as to how man-made objects are instantly distinguishable from naturally occurring ones. Nor do we know if there is value to life because science cannot comment on the value of anything. As real and as intrinsic as meaning, design, and goodness are in everyday experience, science fails to account for them. The scientific explanation of our existence is a sham. We cannot even say if God exists because science cannot test that "hypothesis."

Creation asserts that rationality, intent, meaning, and value preceded nature—that mind preceded matter. Creation thereby explains the rational order of nature, why inference is valid, and why scientific inquiry is worthwhile. Creation explains intervention in the natural order originating from intent and design, why the question, "What is the meaning of this?" is a meaningful question, and why there is value in living. More than just a thought system, creation is reality. The partition between science and religion might enjoy better service if repositioned to separate ideology from fact.

The separation of scientific and religious thought has also paved the way for a Machiavellian stratagem. The First Amendment's protection of the right against an establishment of religion has been used to drive the fact of creation from public consciousness and public institutions. Yet the basis of individual rights, plainly stated in the Declaration of Independence, is that rights are an endowment from the Creator. The Constitution by means of the Bill of Rights goes on to enumerate further rights by specifically prohibiting the government, on the authority of the citizens so endowed, from abridging those rights. While the degree to which the Constitution is linked to the philosophical grounds set forth in the Declaration of Independence may be a matter of debate, the linkage cannot be denied, especially after the Thirteenth, Fourteenth, and Fifteenth amendments. A fact to which the Constitution therefore assents, that human beings are created, cannot be declared unconstitutional. But our federal judges and supreme court justices have barred this fact from public education as an unconstitutional establishment of religion. In so doing, they have acted as agents of perversion, pulling the groundwork out from beneath the Bill of Rights and nullifying the authority to which government must yield. Again, we ask, "Is science being used to serve some other agenda?"

This is not the end of the matter. Science appears to have made the world safe from its Creator, but at one point, the Creator entered the world. The testimony of that entry remains as real a presence in the world as any stone, tree, or other object of scientific inquiry. That testimony is a stand-alone, self-authenticating premise that is neither circular, self-defeating, nor a dead-end.

Appendix A:
Randomness or
Natural Law?

Evolutionary science may be found appealing to randomness (or chance) or to cause-and-effect laws of nature in explanations of how things got to the state they are in today. In discussing evolutionary change such as life origination or the progression of one species to another, one can rely on random mechanisms, or one can appeal to the sufficiency of natural law. Is evolution predictable and repeatable as it should be if operating according to natural law? That is, if you were to start over again with the same environmental factors, would you end up with the same result? Or is it unpredictable and unrepeatable as it would be if operating on randomness or chance, never producing the same result even with all other factors remaining equal? Trying to explain evolutionary change by appealing to both randomness and natural law at the same time for the same process is an offense to reason, though this has been tried.

Natural processes often do conform to lawful regularity and are thus predictable. But some processes are inherently random. A probabilistic analysis becomes necessary to make sense out of such natural phenomena, and this is often successful. But consider DNA, the blueprint of all living things. The number of degrees of variation possible in a DNA molecule is effectively infinite, far too vast to ever be mapped out. How to determine the probability of any given variation is similarly beyond comprehension. To gain an appreciation for what confronts the evolutionary biologist, consider that the number of possible variations—that is, the size of the probability map—for bacterial DNA is approximately $4e4,700,000$.[1] This is an unbelievably huge number. If someone were to write a zero every second, it would take them about eleven and a half days of writing zeros without any lunch breaks to write this number. It is about a million zeros long, a number that surpasses even what the federal government may be capable of generating.

A probability map is what would define the probability of certain evolutionary pathways should the DNA blueprint be modified, such as might occur by a mutation. A probability map would also answer how probable is the survival of any one degree of variation in the DNA blueprint. But a probability map of the magnitude necessary for even bacterial DNA is incalculable, and scientists have no idea of the shape or contours of it. It cannot even be determined if evolutionary pathways exist through the map—that is, if a sequence of successive variations produces viable organisms at each step. In theory, it is possible that some pathways exist, but that cannot be demonstrated. Without knowing the probability map, it is impossible to say for certain how likely the transitions in the fossil record were—that is, whether or not the probabilities favored them. Nor does anyone know if there are forces in nature sufficient to take a DNA blueprint through an evolutionary progression.[2]

But this is just evolutionary transition from one living form to another, a trivial problem compared to the probabilities involved in how life might have started in the first place. The origin of life is an even greater mystery. No one can figure out how DNA, an information-carrying biochemical code, arose from random interactions of molecules. And no one can figure out how both DNA and a functioning entity with a mechanism for replicating DNA, both of which are necessary, came into being simultaneously, together and intact. Not only is DNA itself needed, but the metabolic process running under the direction of the DNA is needed at the same time so that DNA molecules can be replicated. DNA is of no use by itself unless it can be replicated, and the replication process cannot exist without the instructions that DNA supplies. The probabilities against both DNA and a replication process coming together at the same time are stacked so heavily that probability figures simply cannot be used (and are never used) to give the occurrence any credibility. An appeal to randomness without a probability analysis is an admission that evolutionary scientists do not have the slightest idea about what to attribute the existence of living things to.

In essence, natural processes that are characterized as random (typically random collisions of molecules) are being relied upon to explain how successive increases in order are possible. Intuitively, randomness always seems to lead to disorder, not increasing order. The burden of proving otherwise, the need to produce empirical evidence of an instance of disorder to order in biology or biochemistry, is on the evolutionary scientist.[3]

An appeal to natural laws encounters a different, though equally intractable difficulty. There are no natural laws that pertain to evolution because, whether in the laboratory or in the field, neither life origination nor evolutionary progression from one species to another is being observed today. Any appeal to natural law is saddled with the necessity that what happens today happened yesterday and will happen again tomorrow. That is science's assumption of uninterrupted uniformity. So if it does not happen today, it must not have happened yesterday either. They say evolution is occurring too slowly to be observed. Okay, perhaps it is occurring too slowly. But like it or not, the cornerstone of the scientific method is empirical verification. Appeals to randomness and natural law should both be declared dead on arrival.

Appendix B:
Remarks Pertaining to the
Validation of Premises

The premise problem in philosophy may be illustrated by an example often used to exhibit deductive reasoning. Consider the categorical syllogism: "All men are mortal. Socrates is a man. Therefore, Socrates is mortal." The premise is "All men are mortal." It is not a premise that anyone would question, and doubtless, it is accepted by all. However, acceptance is one thing; proof is another. In order to be assured of the reliability of a premise, it is desirable that it be proven in some manner. Initially, there appear to be two means by which that may be achieved: empirically or logically. The premise may be proven empirically by observational verification that indeed all men are mortal. This requires observing the mortality of all men. Or it may be proven logically if it may be deduced from a more general premise that is already proven. For example, all living things are mortal. Men are living things. Therefore, all men are mortal.

Difficulties now arise. How is it possible to observationally verify the mortality of all men? Short of complete verification of all instances of men, the premise remains unproven. Observing ninety-nine out of one hundred mortalities is not sufficient. The inadequacy of the empirical (or inductive) approach is revealed: A universal categorization cannot be proven by a subset of particular instances. Or how is it possible to prove the certainty of an even more general premise? Far from solving the difficulty and proving the original premise, a larger difficulty is introduced. The more general premise, "All living things are mortal," must now be proven. And worse, in the more pressing questions of philosophy, the truth or falseness of premises is not so immediately apparent as "All men are mortal."

The inability to prove a philosophical presupposition (or scientific theory) does not mean it is false. It just means that its status as truth is tentative. The standing of presuppositions and theories tends to be a matter of acceptance rather than proof. Science has had much more success with its theories than philosophy has had with its presuppositions, but the knowledge problem still exists. There are assumptions imbedded even within scientific theories. Assumptions behind numbers of now discarded scientific theories of the past appeared justifiable in their time, but they were later shown to be false or of limited applicability. Assumptions behind today's theories have a similar vulnerability. Pragmatic directions in philosophy and in the philosophy of science grant the uncertainty as permanent, instead emphasizing what works or what produces practical results (sometimes leaving open the question as to which practical results should be produced). But pragmatic inclinations never completely displace the desire for universal certainty and the idea that there is real truth out there to be found; they strangely persist.

We should not give up hope. Nevertheless, in considering the universal character of premises involved in the evolution-creation controversy, it is clear that there can be no premise more universal than an already universal premise; a deductive proof is precluded. And unless one is able to know everything there is to know about everything, empirical proof is beyond reach, especially if that premise involves a negation, such as proving the absence of any naturally originated thing (meaning that, if a supernatural origin is presumed, there is nothing in existence that was not intended) or the absence of any supernaturally originated thing. There are, however, means other than the two identified in the above example to validate or disqualify premises involved in the evolution-creation controversy. These means are forthcoming from the discussions of chapters 9, 10, and 11.

Notes

Introduction

1. "Bible and science teach us different lessons," *The Columbus Dispatch*, June 3, 2007. Used by permission.

2. "Intelligent-design defender has poor grasp of science," *The Columbus Dispatch*, January 21, 2006. Used by permission.

3. "Close, but not quite," *The Columbus Dispatch*, January 22, 2006. Used by permission.

Chapter 1: Science 1

1. Magellan's voyage around the world was not a complete circumnavigation because he was killed in the Philippine Islands. But he had years earlier arrived in the Philippines from the other direction, so he really was the first person to circumnavigate.

2. Renaissance humanism retained the sense of divine presence and gifting of mankind and should not be confused with the secular humanism of the twentieth century built on agnosticism or atheism. See Dale Jorgenson, *Christianity and Humanism* (College Press, 1983), pp. 33, 43.

3. Toby E. Huff, *The Rise of Early Modern Science: Islam, China and the West* (Cambridge University Press, 1993), p. 129–30.

4. John Henry, *The Scientific Revolution and the Origins of Modern Science*, 2nd edition (Palgrave Publishers, 2002), p. 28.

5. Henry, p. 13.

6. Henry, pp. 63–4, 66.

7. Henry, pp. 89–90.

8. Henry, pp. 95–7.

9. A case in point is William Paley's 1802 *Natural Theology; Or Evidences of the Existence and Attributes of the Deity, Collected from the Appearances of Nature.*

10. The telescope was not Galileo's invention, but he improved it and was the first to make discoveries with it. He observed phases of Venus similar to phases of the Moon, rugged features on the surface of the Moon, and four moons circling Jupiter. These observations did not prove heliocentricity, but they weighed in favor of it. Newton's laws of motion and gravitation ultimately established heliocentricity by showing that objects fall to Earth because of the gravitation caused by mass, not because Earth is the center of the universe.

11. Kitty Ferguson, *Measuring the Universe: Our Historic Quest to Chart the Horizons of Space and Time* (Walker and Company, 1999), p. 101.

12. 2 Peter 3:4 RSV.

13. Huff, p. 10.

14. Thomas Aquinas was a leading contributor to the scholastic movement of the late Middle Ages. Scholasticism applied methods of rational argument to inform theology and supplement revelation to produce a synthesis of faith and reason. Aquinas's most important work was *Summa Theologica*, which was written between 1265 and 1274.

15. Huff, pp. 88, 105.

16. Huff, p. 1.

Chapter 2: Genesis

1. A chart showing how this is done may be found in Walt Brown's *In the Beginning*, 7th edition, 2001, p. 272. It is noteworthy that the form of these genealogical records has no parallel in the rest of the Bible. The particular form suggests an invitation to make the calculations.

2. The age of creation was a matter of speculation from time to time prior to the seventeenth century, though it appears that there were no definite answers forthcoming. See Hugh Ross, *A Matter of Days: Resolving a Creation Controversy* (NavPress, 2004), pp. 41–9.

3. Wilbur Applebaum, *The Scientific Revolution and the Foundations of Modern Science* (Greenwood Press, 2005), p. 115.

4. The gap theory was proposed by Thomas Chalmers, a theologian, in 1814 (Brown, p 274). Names and dates were not found for the other theories, but there is a general consensus among at least some authors on this topic that they appeared either as uniformitarian geology was becoming established or at some later time: David L. Wilcox, *God and Evolution: A Faith Based Understanding* (Judson Press, 2004), p. 30; Vernon Blackmore and Andrew Page, *Evolution: The Great Debate* (Lion Publishing, 1989), p. 36; Alan Hayward, *Creation and Evolution: Rethinking the Evidence from Science and the Bible* (Bethany House, 1985), pp. 73, 165–7; Edward J. Larson, *Summer for the Gods: The Scopes Trial and America's Continuing Debate over Science and Religion* (Basic Books, 1997, 2006), pp. 15–6.

5. The Hebrew word *yowm* can mean either a literal day or an indefinite time period (see *Strong's Expanded Exhaustive Concordance of the Bible*, Thomas Nelson Publishers, 2010).

6. See also Psalm 147:8 and 148:4.

7. Psalm 104:5 along with Psalm 93:1 or Psalm 96:10 is sometimes referenced to assert that the Bible says the Earth is stationary and neither rotates on its axis nor revolves around the Sun. However, in the context of the Old Testament, the foundations of the Earth are more generally portrayed as wisdom, knowledge, and righteousness. See Jeremiah 10:12. In Psalm 11, the ordering of the world according to righteousness is the context in which verse 3, referring to the foundations, is placed. This same context is seen in Psalm 82, wherein the Earth's foundations are referenced in verse 5. See also Isaiah 24:18 and context. From time to time, one may also hear or read unsubstantiated claims saying that the Bible says the Earth is flat. Isaiah 40:22 says it is circular. Job 26:10 makes sense only if it is spherical.

8. See also Exodus 16:21–30 and 31:12–7.

9. Some criticisms of Genesis 1 and 2 allege that the second half of Genesis 2:4, "In the day that the Lord God made the earth and the heavens …" begins a second and contradictory account of creation. Alternate sentence demarcations in various translations of verses 4 through 7 may make a definitive conclusion on this matter elusive, but possibilities do exist that the

initial perceptions of an inconsistency are not necessarily sustained. There is the possibility that the term *day* means the sense of *a time* rather than one rotation of the Earth. The Hebrew language term is general enough to permit either sense. An alternate sense of *day* in Genesis 2:4 does not nullify a different meaning of chapter 1's use of the term. Besides the meaning of the word *day*, verses 5 and 6 appear to suggest a sequence that is inconsistent with that of chapter 1. These verses are marked as parenthetical in some translations and a complete sentence would remain if these two verses were skipped. The focal point of the entire sentence from verse 4 through verse 7, therefore, is the creation of man from the ground. The plants and herbs of the field may not necessarily be the same plants and trees that preceded man's creation in chapter 1; they may have been reserved for later appearance. See Genesis 3:18.

10. There is a side question with regard to the Sun, Moon, and stars being created on the fourth day, though this does not affect the answer to the age question. What was the source of light during the first three days? Is the Hebrew terminology flexible enough to translate the Sun, Moon, and stars as having appeared or made visible in the firmament rather than strictly created on the fourth day? Obviously, we know that these exist in outer space, not in the firmament (atmosphere) between the waters. Could they have been part of the heavens created in Genesis 1:1? What else would define the heavens? Otherwise, the light source is unknown. If the Hebrew text could be translated accurately in this manner, then the possibility exists that these objects were brought into existence as the heavens along with the Earth in Genesis 1:1, that the sunlight was permitted to penetrate the cloud cover on the first day, and that the cloud cover dissipated sufficiently that the Sun, Moon, and stars became visible on the fourth day. Another proposal for the source of light during the first three days is the glory of God. But this is probably not a source that would be considered to have been created. The phrase "Let there be—" is used repeatedly throughout chapter 1 in reference to things being created.

11. The term *theistic* is derived from the Greek *Theos*, meaning God. *Theology* is the study of the nature of God. *Atheistic* is the familiar antonym meaning belief in the nonexistence of God.

12. The Genesis 1 creation narrative has correlations with the new creation and the new covenant. In that sense, it is also allegorical. But this point does not constitute grounds for dismissing the historical character of the narrative. See Galatians 4:21–31 for a reference to historical records in a similar allegorical sense.

Chapter 3: Geology

1. Vernon Blackmore and Andrew Page, *Evolution: The Great Debate* (Lion Publishing, 1989), p. 29.

2. Intuitively, a single flood would not seem to have resulted in distinguishable layers. On the other hand, a flood of worldwide scope is difficult to conceive and may have had an entirely different character. The reasoning for concluding that each layer was a separate event, while persuasive, may not constitute a certain dismissal of one flood being responsible for the majority of formations. In any case, uniformitarians have a similar problem in explaining what brought a stop to one layer and what resulted in initiation of the next.

3. In 1778, George-Louis Leclerc Comte de Buffon said that the earth's topographical nature could be explained as the result of present processes, but he did not develop this thought further.

4. Research & Education Association, *Super Review of Geology*, Carl Fuchs, editor (Research & Education Association, 2003), pp. 4, 7, 32, 34–5, 38, 168–9.

5. Hugh Ross, *A Matter of Days: Resolving a Creation Controversy* (NavPress, 2004), p. 176.

6. Research & Education Association, p. 19.

7. Research & Education Association, p. 19.

8. Research & Education Association, p. 20.

9. Research & Education Association, pp. 17–8, 42.

10. The Institute for Creation Research Radio-Isotopes and the Age of the Earth (RATE) Project has examined this and certain other lines of evidence that radioactive decay rates have been faster in the past.

11. The necessity of rapid burial in fossil formation is mentioned in the following textbooks: Frederick K. Lutgens and Edward J. Tarbuck, *Essentials of Geology*, 6th edition (Prentice Hall, 1998), p. 379; Frederick K. Lutgens and Edward J. Tarbuck, *Foundations of Earth Science*, 6th edition (Prentice-Hall, 2011), p. 248; Stephen Marshak, *Essentials of Geology*, 3rd edition (W. W. Norton & Company, 2009), p. 270; David McConnell, David Steer, Catharine Knight, and Katharine Owens, *The Good Earth*, 2nd edition (McGraw-Hill, 2010), p. 220. These textbooks describe the requirements

of fossil formation separately from descriptions of the formation of fossil bearing strata. Other textbooks do not mention rapid burial: W. Kenneth Hamblin and Eric H. Christiansen, *Earth's Dynamic Systems*, 10th edition (Prentice Hall, 2004); Stanley Chernicoff, Haydn A. Fox, and Ramesh Venkatakrishnan, *Essentials of Geology* (Worth Publishers, 1997). *Essentials of Geology* by Marshak suggests that fossilization is not generally occurring today (p. 270).

12. John Baumgardner, "Carbon Dating Undercuts Evolution's Long Ages," Institute for Creation Research, www.icr.org, 2003; Larry Vardiman, "Diamonds May be a Creationist's Best Friend," Institute for Creation Research, www.icr.org, 2008. Diamonds are thought to have formed from organic carbon that was subducted into the crust of the earth, formed into diamonds, and then later brought to the surface by volcanic activity. The geology textbooks listed in note 11 are silent on the presence of C-14 in coal and diamonds.

13. The significance or insignificance of discontinuities in the fossil record, so often the center of the evolution-creation debate, is moot if the entire fossil record was deposited at once. In that case, the fossil record simply consists of all the organisms existing at the time of the flood that were prone to burial.

14. Walt Brown, *In the Beginning*, 7th edition, 2001, pp. 31–31, 80.

Chapter 4: Biology

1. See Vernon Blackmore and Andrew Page, *Evolution: The Great Debate* (Lion Publishing, 1989), p. 10.

2. Blackmore and Page, p. 48.

3. The complete title is "On the Origin of Species by Means of Natural Selection, or the Preservation of Favoured Races in the Struggle for Life."

4. The classification scale of living things ordered from the most general grouping to the most specific is the following: phyla, class, order, family, genus, and species.

5. Thomas Henry Huxley, *On the Reception of the 'Origin of Species,'* Quarterly Review, 1887; reprinted in: *The Life and Letters of Charles Darwin*, Vol. 1, Francis Darwin, editor (Appleton, 1904).

6. The Humanist Manifesto of 1933 and later the Humanist Manifesto II of 1973. See also Dale Jorgenson, *Christianity and Humanism* (College Press, 1983), pp. 55–6.

7. Rather than preventing a federal government sponsored church as the phrase was originally intended by Thomas Jefferson in commenting on the First Amendment, the courts now construe this as meaning that churches cannot influence affairs of government, including government-run education.

8. Proverbs 29:18 NIV: "Where there is no revelation, the people cast off restraint."

9. Peter J. Bowler, *Evolution: The History of an Idea* (University of California Press, 2003), p. 143. Michael Ruse's book is titled *Monad to Man: the Concept of Progress in Evolutionary Biology.*

Chapter 5: Astronomy and Physics

1. Aristarchus observed that the length of time that the Moon was completely hidden during a total lunar eclipse was about equal to the time between when the Earth's shadow just begins to obscure the Moon and the Moon becomes completely hidden. He concluded from this that the diameter of the Earth's shadow where the Moon crosses it must be about twice the Moon's diameter. He knew from solar eclipses that the angular sizes of the Sun and Moon as viewed on Earth are the same and that the Sun is further away than the Moon. He assumed that the Sun is large enough and far enough away that the angular size of the Earth's shadow would have the same angular size as the Sun. Then, with an approximate measurement of the Sun's (or Moon's) angular size as viewed on Earth and by straightforward geometry, he deduced the Moon's size and distance. He correctly reasoned that the Moon's diameter must be about one fourth the diameter of the Earth and its distance about sixty times the Earth's radius. See Kitty Ferguson, *Measuring the Universe: Our Historic Quest to Chart the Horizons of Space and Time* (Walker and Company, 1999), p. 25.

2. John Gribbin, *The Birth of Time: How Astronomers Measured the Age of the Universe* (Yale University Press, 1999), p. 74.

3. A light-year is the distance light traverses in one year, 9,460,000,000,000 kilometers or 5,865,000,000,000 miles.

4. Ferguson, p. 223.

5. Richard Wolfson, *Simply Einstein: Relativity Demystified* (W. W. Norton, 2003), p. 164.

6. Simon Singh, *Big Bang: The Origin of the Universe* (HarperCollins, 2004), p. 101–2.

7. Wolfson, p. 159.

8. Wolfson, p. 216.

9. Wolfson, p. 216.

10. Singh, p. 135–6.

11. Gribbin, p. 93–4.

12. Ferguson, p. 224–5.

13. Job 9:8; Psalm 104:2; Isaiah 40:22, 42:5, 44:24, 48:13; Jeremiah 51:15; Zechariah 12:1.

14. Robert C. Newman, *Genesis One and the Origin of the Earth* (InterVarsity Press, 1977), pp. 16, 21.

15. In view of Jeremiah 31:37, it may even be unscriptural.

16. Newman, p. 21.

17. Newman, p. 28.

18. Genesis 18:14, Jeremiah 32:17, 27 RSV.

19. Russell Humphreys, *Starlight and Time: Solving the Puzzle of Distant Starlight in a Young Universe* (Master Books, 1994).

20. Humphreys links the chronology of his theory to Genesis 1 days in a particular fashion that does not agree that well with the interpretation of Genesis 1 described in chapter 2 of this book. I would link all his theory's chronology to Genesis 1:1 only.

Chapter 6: Science 2

1. C. S. Lewis, *Miracles: A Preliminary Study* (HarperSanFrancisco, 2001), p. 2. MIRACLES by C. S. Lewis, copyright C. S. Lewis Pte. Ltd. 1947, 1960. Extract reprinted by permission.

2. Donald Goldsmith, *Einstein's Greatest Blunder? The Cosmological Constant and Other Fudge Factors in the Physics of the Universe* (Harvard University Press, 1995), p. 5.

3. Trudy E. Bell, Dave Dooling, and Janie Foulke, "Threshold of the New Millennium" (From an interview with physicist Arno A. Penzias), *IEEE Spectrum*, October 1999, p. 63. Used by permission.

4. The if-then syllogism typifies scientific reasoning: *If* a certain hypothesis is true, *then* certain results should follow. It is tempting to regard the occurrence of the expected results as proving the hypothesis. In formal logic, this is known as the error of *affirming the consequent*. Much of the painstaking work in science involves isolating the factors that cause the expected results in order to eliminate the possibility that other hypotheses could be the cause of the expected results, thus minimizing the logical error.

5. For those interested in continuing investigation of this point, Walt Brown's *In the Beginning* (http://www.creationscience.com/onlinebook/index.html) contains many additional scientific findings.

6. Bell, *et al.*, p. 63. Used by permission.

Chapter 7: Controversy

1. The Roman Catholic position toward evolution was one of cautious acceptance. Theistic evolution was at first considered to be not incompatible with Catholic theology, but when evolution became less compatible with theism, scientific and theological pursuits were strictly segregated. Historical criticism and modernism were officially condemned in 1907. As a whole, Roman Catholicism did not contribute much to the reemerging evolution-creation controversy of the 1920s or thereafter. The church also had its own schools.

2. The ACLU was formed during WWI to defend conscientious objectors and antiwar protesters. After the war, it focused on defense of labor leaders and thereafter pursued defense of academic freedom and free speech. See Edward J. Larson, *Summer for the Gods: The Scopes Trial and America's Continuing Debate over Science and Religion*, p. 61.

3. Edward J. Larson, *Summer for the Gods: The Scopes Trial and America's Continuing Debate over Science and Religion* (Basic Books, 1997, 2006), p. 50. The Butler Act read in part: "Be it enacted by the General Assembly of

the State of Tennessee, that it shall be unlawful for any teacher in any of the Universities, Normals and all other public schools of the State which are supported in whole or in part by the public school funds of the State, to teach any theory that denies the story of the Divine Creation of man as taught in the Bible, and to teach instead that man has descended from a lower order of animals."

4. Edward Caudill, *et. al.*, *The Scopes Trial: A Photographic History* (University of Tennessee Press, 2000), p. 7. Larson, pp. 100–1.

5. Caudill, p. 33.

6. Another indication of the surprise nature of this move was that most of the reporters, assuming that remaining trial proceedings would be anticlimactic, had either left town over the weekend or were loafing. Scopes said he was even recruited to write stories for some of them. See R. M. Cornelius, *World's Most Famous Court Trial* (http://www.bryan.edu/802.html).

7. R. M. Cornelius, *William Jennings Bryan, The Scopes Trial, and Inherit the Wind* (Bryan College, 2004), p. 3.

8. Caudill, p. 18. Larson, pp. 220–1.

9. Larson, pp. 224–5.

10. Larson, pp. 230–1.

11. Caudill, pp. 8, 70–1. Larson, pp. 225, 244–5.

12. "Upon the principal issue, that of constitutionality, the court holds that Initiated [statute] is a valid exercise of the state's power to specify the curriculum in its public schools. The court expresses no opinion on the question whether the Act prohibits any explanation of the theory of evolution or merely prohibits teaching that the theory is true; the answer not being necessary to a decision in the case, and the issue not having been raised."

13. Edward J. Larson, *Trial and Error: The American Controversy Over Creation and Evolution*, 3rd edition (Oxford University Press, 2003), p 185.

14. In the case of Oklahoma, the antievolution statute was an amendment to a 1923 textbook bill that was repealed in 1925 for reasons unrelated to

the evolution issue. In view of the *Epperson* decision, the Mississippi statute was overturned by that state's supreme court in 1970.

15. Criteria specified in the *Lemon v. Kurtzman* 1971 US Supreme Court opinion have been used since 1971 for judging constitutionality under the Establishment Clause of the First Amendment. These criteria are three: (1) a statute must have a secular purpose, (2) its effect must not be to favor or disfavor a religion, and (3) the degree of entanglement of government in religion must not be "excessive."

16. Justice Scalia's dissenting opinion in this case argued that the academic freedom intent of the legislation should rather have been construed as freedom from indoctrination, the academic freedom of the students to be presented with alternative views, not necessarily that of the teachers to teach what they would. I agree with this argument but would offer a qualification: Given the types of social influences on students in high school and the level of maturity, I do not see how they could possibly be in a position to choose for themselves on such a contentious and intellectually demanding subject as evolution-creation. To a large extent, education through the twelfth grade *is* indoctrination. I do not see how the teacher-student relation could be viewed as anything approaching equality.

17. Phillip E. Johnson, *Darwin on Trial*, 2nd edition (InterVarsity Press, 1993), p. 69.

18. Johnson, p. 158.

19. Michael Behe, *Darwin's Black Box: The Biochemical Challenge to Evolution* (Simon & Schuster, 1998).

20. Eugenie C. Scott, *Evolution vs. Creationism: An Introduction* (Greenwood Press, 2004), p. 117.

21. If evolutionary theory and the particular style of humanism it undergirds represent an ideological position (and they do), separation of church and state is not really what is being accomplished here. Moreover, an ideologically neutral education may well be an impossibility. What these organizations suppress and what many fail to recognize is that the US Constitution and the Bill of Rights are built on an ideology opposite to the one now inserted into public education. Individual rights are not sustainable apart from a general acknowledgment of their transcendent source; everyone knows what atheistic political ideologies do to individual rights. It is a rather odd state of affairs that the First Amendment should now be made to sustain an ideology hostile to rights and to prohibit the one that lays a foundation for them.

22. The idea that God used evolution as the means of creation (theistic evolution) is on tenuous ground because (1) to scientists, the theistic aspect is superfluous and (2) an honest interpretation of the Genesis record indicates that God did not use incremental development over long periods of time in creating. Scientists have no use for the theistic aspect; God disavows the evolutionary aspect. What remains? See also chapter 2 of this book.

23. An example of a sub-theory would be "punctuated equilibrium," an alternate theoretical concept of biological evolution presupposing that long periods of no evolutionary change, or equilibrium were punctuated by short periods of rapid evolutionary change. This sub-theory, originated by Niles Eldridge and Stephen Jay Gould, was devised to better correlate evolutionary progression with the fossil record. It stands in contrast to the standard evolutionary concept of continuous and slow incremental change.

24. Judge William Overton, in his opinion on *McLean v Arkansas*, referred to a "contrived dualism" in noting that evidence against evolution presented by creation science was the evidence used to validate creation science (Section IV.(A)). His objection to this strategy in principle is correct because other alternatives could exist. But in the case of evolutionary science and creation science, no other alternatives exist. Either life was caused to exist by a creator, or it came to exist on its own. An attempt at a third alternative interjected during this trial was that life somehow arrived from outer space, but that just defers the question. What was responsible for it being formed somewhere in outer space? Judge Overton also drew a distinction between a narrow definition of evolution claimed by scientists as development of species and a wider definition of evolution as the origin of life as understood by the creation scientists. This argument was advanced by the plaintiffs to say that evolution does not refer to the origination of life, only its subsequent development. The strategy here was to imply that evolution and creation are not two answers to the same question. But this suppresses the logical implications of evolutionary theory obvious from the time it was first proposed by Darwin, even to the nonscientist.

25. Larry A. Witham, *Where Darwin Meets the Bible: Creationists and Evolutionists in America* (Oxford University Press, 2002), p. 71, from an interview with and quoting Eugenie C. Scott, Director of the National Center for Science Education (NCSE).

26. Edward J. Larson in Afterword to *Summer for the Gods*, p. 274, references this quote from a 1998 US National Academy of Sciences position statement.

27. Phillip E. Johnson in *The Right Questions: Truth, Meaning and Public Debate* (InterVarsity Press, 2002), p. 44, references this quote from a 1981 US National Academy of Sciences position statement.

28. Values and ethics are often claimed to be subjective preferences, not objective truths. Under this presumption, they can be explained as the result of material causes, and thus, they are potentially explainable by science. The error in this thinking is a subject that will be addressed in later chapters.

29. Analogy drawn from C. S. Lewis's *Miracles: A Preliminary Study* (HarperSanFrancisco, 2001), pp. 151–8.

Chapter 8: Thought

1. Alternately, two contradictory statements cannot both be true.

2. C. S. Lewis, *Miracles: A Preliminary Study* (HarperSanFrancisco, 2001), pp. 20, 26. MIRACLES by C. S. Lewis, copyright C. S. Lewis Pte. Ltd. 1947, 1960. Extract reprinted by permission.

3. The value of truth really is accepted for its own sake irrespective of any pragmatic survival benefit. The two criteria reside in different realms.

4. Lewis, "De Futilitate," *Christian Reflections*, Walter Hooper, editor (William B. Eerdmans, 1967), p. 64, emphasis in original.

5. Of course, if particular knowledge were to be found false at a point, it would cease to be regarded as knowledge.

6. Lewis, *Miracles*, p. 65. MIRACLES by C. S. Lewis, copyright C. S. Lewis Pte. Ltd. 1947, 1960. Extract reprinted by permission.

7. Lewis, *Miracles*, pp. 38–39. MIRACLES by C. S. Lewis, copyright C. S. Lewis Pte. Ltd. 1947, 1960. Extract reprinted by permission.

Chapter 9: Philosophy 1

1. Colossians 2:8 RSV.

2. Ravi Zacharias, *Deliver Us From Evil* (Word Publishing, 1996) p. 219.

3. Immanuel Kant (1724–1804), regarded as one of the most eminent philosophers in the Western intellectual tradition, introduced this problem to the field of epistemology in his work *The Critique of Pure Reason*. It was actually part of the solution to a problem of skepticism posed by some earlier philosophers, but that is a different story.

4. David L. Wilcox, *God and Evolution: A Faith Based Understanding* (Judson Press, 2004), pp. 11–2.

5. George Gaylord Simpson, *The Meaning of Evolution*, revised edition (Yale University Press, 1967), pp. 344–5.

6. C. S. Lewis's 1943 essay, "The Poison of Subjectivism," explains the contradictions of subjectivism (here synonymous with relativism) and how games are played with values under the pretense of "improving" them perhaps better than any other work. See C. S. Lewis, *Christian Reflections*, Walter Hooper, editor (William B. Eerdmans, 1967), pp. 72–81.

7. This example and those that follow in this section summarize some of the self-defeating philosophical premises discussed by Ravi Zacharias in *Can Man Live Without God* (Word Publishing, 1994), pp. 195–205.

8. C. S. Lewis, *Miracles: A Preliminary Study* (HarperSanFrancisco, 2001), pp. 17–36.

Chapter 10: Philosophy 2

1. Colossians 2:3.

2. Note that miracles, often thought of as violations of laws of nature, can also be thought of as the same sort of intrusions into nature. Miracles do not need to be construed as violations of laws of nature. See C. S. Lewis, *Miracles: A Preliminary Study* (HarperSanFrancisco, 2001), pp. 87–98.

3. Besides naturalism, another position that denies free will (though preserving intent) is the religious doctrine known as Calvinism (after the sixteenth century theologian). In Calvinism, intent originates from God. This doctrine similarly falls by the demand for personal rather than theological answers.

4. To clarify the linkage between values and good and evil by example: Freedom is valued, truth is valued, and life is valued. Assuming no conflict between these values, any action that respects, preserves, elevates, clarifies,

or enhances these values is good. Any action that diminishes, denies, confuses, or destroys them is evil.

5. A question may be raised here about God commanding the Israelites in the Old Testament to annihilate the nations occupying the promised land before them. This is a theological question beyond the scope of this book. But I would venture that there are certain facts relevant to any answer: (1) God owns life, people do not; (2) even after removal of much of the idolatrous influence of those nations and repeated warnings of the prophets, it was still about nine hundred years before the Israelites began to consistently observe even the first two of the Ten Commandments; (3) more than once during that nine hundred years, other nations were sent to nearly wipe out the Israelites for that very reason.

6. Genesis 1:26–7.

7. Exodus 20:13–17.

8. Deuteronomy 6:4–5, Leviticus 19:18.

9. As an example of conflicting principles, take the *Scopes* trial question of the academic freedom of teachers versus the right of the taxpaying public to weigh in on the content of state funded education. Neither of these was ever questioned as a worthy principle, but they were in serious conflict.

10. C. S. Lewis, "De Futilitate," *Christian Reflections*, Walter Hooper, editor (William B. Eerdmans, 1967), p. 70.

11. I take that back; we cannot bypass the question completely, seeing how meaning is central to the dispute between evolution and creation. A lead-in to an answer might be put this way: The privilege of free will is in the face of the often consequential and serious nature of decisions in this world. But free will does not truly exist unless decisions are consequential. Without a real possibility for evil, free will is not consequential, and without consequential free will, there can be no love.

12. C. S. Lewis, *Miracles: A Preliminary Study*, pp. 21–2. MIRACLES by C. S. Lewis, copyright C. S. Lewis Pte. Ltd. 1947, 1960. Extract reprinted by permission.

13. Awareness of thought raises the question of awareness overall: How is consciousness to be explained scientifically? Awareness and consciousness raise the same sort of questions as objectivity, design, and value, questions for which naturalism can supply no answers but supernaturalism answers straight away.

Chapter 11: Revelation

1. How the need to ground these claims seems so little appreciated is astonishing. But then adverse consequences of departing from established belief systems can be huge. It is also worth noting that politics, because it is effectively religion applied to questions of public interest, often produces the same sort of divisiveness as religion itself.

2. For an introductory study on the specifics of the qualifications of divine authorship, see Jay Wilson's *Proof that the Bible Is the Word of God* (http://www.newcreationstudies.org/newcreation/proof.htm).

3. Psalm 16:8–11, Psalm 22:14–22, and Isaiah 52:13–53:12 contain the chief predictions of the resurrection. See also Acts 2:22–36, 3:12–26, 4:8–12, 5:27–32, 13:26–41, 17:2–3, and 1 Corinthians 15:1–20.

4. Psalm 118:22–3 KJV.

5. C. S. Lewis, *Miracles: A Preliminary Study* (HarperSanFrancisco, 2001), p. 174. Skeptics have tried to explain away the resurrection of Jesus by claiming that (1) it was all just a myth or a legend and he never existed in the first place, (2) he did not really die on the cross and revived later on, (3) the disciples went to the wrong tomb, or (4) the disciples stole the body, hid it, then lied about his resurrection. The independent and contemporaneous accounts of the New Testament continue to stand against unverifiable guesses of later centuries. The best that the leaders of the Jews could do at the time was to pay "keep quiet" money to the guards who were stationed at the tomb for the express purpose of preventing the disciples from stealing the body, telling them to say the disciples stole the body while they were sleeping (see Matthew 28:11–5).

Chapter 12: Reason and Faith

1. Psalm 14:1, Proverbs 26:12.

2. We need to keep in mind, of course, that the "knowledge" that knowledge can only be obtained by scientific methods is not knowledge obtained by scientific methods.

3. Phillip Johnson, in *Defeating Darwinism by Opening Minds* (InterVarsity Press, 1997), describes what is happening with the separation of faith and reason in education and in the culture at large, the extreme cultural bias that favors philosophical naturalism, and the difficulties faced in overcoming

that bias. Ravi Zacharias, in *Deliver Us From Evil* (Word Publishing, 1996), presents a concise historical perspective on how the current situation of a secular and pluralistic culture wherein all faith is ultimately privatized—that is, excluded from public expression, arose, and how that is an utterly incoherent philosophy for the individual.

4. 2 Corinthians 5:7.

5. Acts 2:14–36, 3:17–26, 4:8–12, 7:2–53, 8:26–35, 13:16–41, 15:13–18, 17:2–3, 26:22–23, 28:23–28. 2 Peter 1:19 and context confirms the point that the words of the Old Testament prophets were regarded as more certain than the apostles' own eyewitness testimony.

6. Hebrews 11:6 RSV.

7. Hebrews 11:1 RSV.

8. C. S. Lewis, "Religion: Reality or Substitute?," *Christian Reflections*, Walter Hooper, editor (William B. Eerdmans Publishing, 1967), p. 42.

9. John 1:1–3, 14, 1 John 1:1–3.

10. Lewis, "The Poison of Subjectivism," *Christian Reflections*, p. 73.

11. Lewis, "Religion: Reality or Substitute?" *Christian Reflections*, p. 43.

12. Ravi Zacharias, *Jesus Among Other Gods*, (W Publishing Group, 2000), p. 60. Reprinted by permission of Thomas Nelson, Inc., Nashville, Tennessee. All rights reserved.

13. Matthew 18:3, Luke 18:17.

14. Matthew 12:38–9 KJV.

15. Zacharias, p. 65.

16. Ravi Zacharias, *Biblical Authority and Our Cultural Crisis, Part I: Cultural Relativism and the Emasculation of Truth*, DVD, Ravi Zacharias International Ministries, 1993.

Appendix A: Randomness or Natural Law?

1. David L. Wilcox, *God and Evolution: A Faith Based Understanding* (Judson Press, 2004), p. 60.

2. Wilcox, p. 60.

3. A common creationist argument appeals to the laws of thermodynamics to make a universal case against disorder to order. The laws of thermodynamics were originally devised to explain the operation of steam engines. They can also be applied to steam cycles in power plants, etc. To apply the laws of thermodynamics to a wider system requires knowing whether or not that system is a closed or open system. Unless one is familiar with these laws as a practicing scientist or engineer, my advice is to steer clear of thermodynamics. Why take the burden of proof away from the evolutionary scientist and place it on yourself when you do not need to?

Bibliography

Applebaum, Wilbur. *The Scientific Revolution and the Foundations of Modern Science*. Westport, CT: Greenwood Press, 2005.

Baumgardner, John. *Carbon Dating Undercuts Evolution's Long Ages*. Institute for Creation Research, www.icr.org, 2003.

Behe, Michael. *Darwin's Black Box: The Biochemical Challenge to Evolution*. New York: Simon & Schuster, 1998.

Bell, Trudy E., Dave Dooling, and Janie Foulke. "Threshold of the New Millennium." *IEEE Spectrum* (October 1999).

Blackmore, Vernon, and Andrew Page. *Evolution: The Great Debate*. Oxford: Lion Publishing, 1989.

Bowler, Peter J. *Evolution: The History of an Idea*. Berkeley: University of California Press, 2003.

Brown, Walt. *In The Beginning: Compelling Evidence for Creation and the Flood*, 7th edition, Center for Scientific Creation, 2001.

Caudill, Edward, Edward Larson, and Jesse Fox Mayshark. *The Scopes Trial: A Photographic History*. Knoxville: University of Tennessee Press, 2000.

Chernicoff, Stanley, Haydn A. Fox, and Ramesh Venkatakrishnan. *Essentials of Geology*. New York: Worth Publishers, 1997.

Cornelius, R. M. *William Jennings Bryan, The Scopes Trial, and Inherit the Wind*. Dayton, TN: Bryan College, 2004.

Ferguson, Kitty. *Measuring the Universe: Our Historic Quest to Chart the Horizons of Space and Time*. New York: Walker and Company, 1999.

Goldsmith, Donald. *Einstein's Greatest Blunder? The Cosmological Constant and Other Fudge Factors in the Physics of the Universe*. Cambridge, MA: Harvard University Press, 1995.

Gribbin, John. *The Birth of Time: How Astronomers Measured the Age of the Universe*. New Haven, CT: Yale University Press, 1999.

Hamblin, W. Kenneth, and Eric H. Christiansen. *Earth's Dynamic Systems*, 10th edition. Upper Saddle River, NJ: Prentice-Hall, 2004.

Hayworth, Alan. *Creation and Evolution: Rethinking the Evidence from Science and the Bible*. New York: Bethany House Publishers, 1985.

Henry, John. *The Scientific Revolution and the Origins of Modern Science*, 2nd edition. Basingstoke, Hampshire: Palgrave Publishers, 2002.

Huff, Toby E. *The Rise of Early Modern Science: Islam, China, and the West*. Cambridge: Cambridge University Press, 1993.

Humphreys, Russell. *Starlight and Time: Solving the Puzzle of Distant Starlight in a Young Universe*. Green Forest, AR: Master Books, 1994.

Johnson, Phillip E. *Darwin on Trial*, 2nd edition. Downers Grove, IL: InterVarsity Press, 1993.

Johnson, Phillip E. *Defeating Darwinism by Opening Minds*. Downers Grove, IL: InterVarsity Press, 1997.

Johnson, Phillip E. *The Right Questions: Truth, Meaning and Public Debate.* Downers Grove, IL: InterVarsity Press, 2002.

Jorgenson, Dale. *Christianity and Humanism.* Joplin, MO: College Press Publishing Company, 1983.

Larson, Edward J. *Summer for the Gods: The Scopes Trial and America's Continuing Debate over Science and Religion.* New York: HarperCollins Publishers, 1997.

Larson, Edward J. *Trial and Error: The American Controversy Over Creation and Evolution*, 3rd edition. Oxford: Oxford University Press, 2003.

Lewis, C. S. *Christian Reflections.* Edited by Walter Hooper. Grand Rapids, MI: William B. Eerdmans Publishing, 1967.

Lewis, C. S. *Miracles: A Preliminary Study.* New York: HarperCollins, 2001.

Lutgens, Frederick K., and Edward J. Tarbuck. *Essentials of Geology*, 6th edition. Upper Saddle River, NJ: Prentice-Hall, 1998.

Lutgens, Frederick K., and Edward J. Tarbuck. *Foundations of Earth Science*, 6th edition. Upper Saddle River, NJ: Prentice-Hall, 2011.

Marshak, Stephen. *Essentials of Geology*, 3rd edition. New York: W. W. Norton & Company, 2009.

McConnell, David, David Steer, Catharine Knight, and Katharine Owens. *The Good Earth*, 2nd edition. New York: McGraw-Hill, 2010.

Newman, Robert C. *Genesis One and the Origin of the Earth.* Downers Grove, IL: InterVarsity Press, 1977.

Research & Education Association. *Super Review of Geology.* Edited by Carl Fuchs. Township, NJ: Research & Education Association, 2003.

Ross, Hugh. *A Matter of Days: Resolving A Creation Controversy.* Colorado Springs, CO: NavPress, 2004.

Ruse, Michael. *Monad to Man: The Concept of Progress in Evolutionary Biology*. Cambridge, MA: Harvard University Press, 1996.

Scott, Eugenie, C. *Evolution vs. Creationism: An Introduction*. Westport, CT: Greenwood Press, 2004.

Simpson, George Gaylord. *The Meaning of Evolution*, revised edition. New Haven, CT: Yale University Press, 1967.

Singh, Simon. *Big Bang: The Origin of the Universe*. New York: HarperCollins Publishers, 2004.

Vardiman, Larry. *Diamonds May be a Creationist's Best Friend*. Institute for Creation Research, www.icr.org, 2008.

Whitcomb, John C., and Henry M. Morris. *The Genesis Flood: The Biblical Record and its Scientific Implications*. Phillipsburg, NJ: Presbyterian and Reformed Publishing, 1961.

Wilson, Jay, *Proof that the Bible Is the Word of God*. Bozeman, MT: 11th Hour Press, 2002.

Witham, Larry A. *Where Darwin Meets the Bible: Creationists and Evolutionists in America*. Oxford: Oxford University Press, 2002.

Wilcox, David L. *God and Evolution: A Faith Based Understanding*. Valley Forge, PA: Judson Press, 2004.

Wolfson, Richard. *Simply Einstein: Relativity Demystified*. New York: W. W. Norton, 2003.

Zacharias, Ravi. *Biblical Authority and Our Cultural Crisis, Part I: Cultural Relativism and the Emasculation of Truth*, DVD. Atlanta, GA: Ravi Zacharias International Ministries, 1993.

Zacharias, Ravi. *Can Man Live Without God*. Nashville, TN: Word Publishing (Thomas Nelson Publishers), 1994.

Zacharias, Ravi. *Deliver Us From Evil*. Nashville, TN: Word Publishing (Thomas Nelson Publishers), 1996.

Zacharias, Ravi. *Jesus Among Other Gods*. Nashville, TN: W Publishing Group (Thomas Nelson Publishers), 2000.

Printed in the United States
By Bookmasters